New Materials

TOWARDS A HISTORY
OF CONSISTENCY

Edited by

Amy E. Slaton

LEVER
PRESS

Copyright © 2020 by Amy E. Slaton

Lever Press (leverpress.org) is a publisher of pathbreaking scholarship. Supported by a consortium of liberal arts institutions focused on, and renowned for, excellence in both research and teaching, our press is grounded on three essential commitments: to publish rich media digital books simultaneously available in print, to be a peer-reviewed, open access press that charges no fees to either authors or their institutions, and to be a press aligned with the ethos and mission of liberal arts colleges.

This work is licensed under the Creative Commons Attribution-NonCommercial 4.0 International License. To view a copy of this license, visit http://creativecommons.org/licenses/by-nc/4.0/ or send a letter to Creative Commons, PO Box 1866, Mountain View, CA 94042, USA.

The complete manuscript of this work was subjected to a partly closed ("single blind") review process. For more information, please see our Peer Review Commitments and Guidelines at https://www.leverpress.org/peerreview

DOI: https://doi.org/10.3998/mpub.11649332
Print ISBN: 978-1-64315-013-0
Open access ISBN: 978-1-64315-014-7

Library of Congress Control Number: 2019954802

Published in the United States of America by Lever Press, in partnership with Amherst College Press and Michigan Publishing

Contents

Member Institution Acknowledgments   v

Chapter One: Introduction   1
   Amy E. Slaton

PART I: MATERIALS TESTED, SUCCESS DEFINED

Chapter Two: Muddy to Clean: The Farm-Raised Catfish Industry, Agricultural Science, and Food Technologies   39
   Karen Senaga

Chapter Three: Room at the Bottom: The Technobureaucratic Space of Gold Nanoparticle Reference Material   73
   Sharon Tsai-hsuan Ku

PART II: MATERIALS PRODUCED, LABOR DIRECTED

Chapter Four: The Scientific Co-Op: Cloning Oranges and Democracy in the Progressive Era   119
   Tiago Saraiva

Chapter Five: Plyscrapers, Gluescrapers, and Mother Nature's Fingerprints   151
   Scott Knowles and Jose Torero

PART III: MATERIALS INTERPRETED, COMMUNITIES DESIGNED

Chapter Six: The Inmate's Window: Iron, Innovation, and the Secure Asylum — 177
Darin Hayton

Chapter Seven: Cultural Frames: Carbon-Fiber-Reinforced Polymers, Taiwanese Manufacturing, and National Identity in the Cycling Industry — 203
Patryk Wasiak

Chapter Eight: Grenfell Cloth — 237
Rafico Ruiz

Afterword: Old Materials — 271
Projit Bihari Mukharji

List of Contributors — 279

Acknowledgments — 283

Index — 285

Member Institution Acknowledgments

Lever Press is a joint venture. This work was made possible by the generous support of Lever Press member libraries from the following institutions:

Adrian College
Agnes Scott College
Allegheny College
Amherst College
Bard College
Berea College
Bowdoin College
Carleton College
Claremont Graduate
  University
Claremont McKenna College
Clark Atlanta University
Coe College
College of Saint Benedict /
  Saint John's University
The College of Wooster
Denison University
DePauw University
Earlham College
Furman University
Grinnell College
Hamilton College
Harvey Mudd College
Haverford College
Hollins University
Keck Graduate Institute
Kenyon College
Knox College
Lafayette College Library
Lake Forest College
Macalester College
Middlebury College

Morehouse College
Oberlin College
Pitzer College
Pomona College
Rollins College
Santa Clara University
Scripps College
Sewanee: The University of the South
Skidmore College
Smith College
Spelman College
St. Lawrence University
St. Olaf College
Susquehanna University
Swarthmore College
Trinity University
Union College
University of Puget Sound
Ursinus College
Vassar College
Washington and Lee University
Whitman College
Willamette University
Williams College

CHAPTER ONE

INTRODUCTION

Amy E. Slaton

We live in the time of nanopants. New, fluid-repellent textiles based on molecular-level innovations in fabric design have brought about pants that resist stains and require less frequent washing than conventional garments. Antibacterial nanosocks, highly durable nanopaints and newly efficacious nanopharmaceuticals have also reached markets over the last decade, each bringing some perceived enhancement in product performance. Each commodity derives from manufacturers' shifting investments, with resources directed toward the inventive efforts of scientists and the subsequent production of goods for consumers in pursuit of novel goods. The result: with sufficient cash in hand, an unprecedented form of pants—or socks, paint, or drugs—can today be yours.

And yet, arguably as much is old in this landscape of innovation as is new. Scientists have been refining the raw materials used by industries for many generations, and markets have thrived on the exchange of novel goods throughout human history; industrial scientists' very choices regarding which topics merit research support long-established distributions of social and economic influence. The uncertain environmental and health consequences of nanoscale manufacturing operations for workers, downstream communities, and consumers echo the risks of countless previous industrial production processes and follow similar patterns of class, race, or other social demarcations. Mass-produced trousers have existed since the mid-nineteenth century, generated through extremely robust patterns of capital-labor relations. Even stains have a history, their stigmatization expressing prevailing standards of hygiene and self-discipline and, in many cultures, class constructs. So perhaps we might ask under what circumstances, exactly, nanopants, reflecting so many institutional and cultural continuities, can come to seem new at all. What's more, the results of "successful" materials research and development may extend widely to alter global living conditions but alternatively—if designed or priced for narrow distribution, selectively marketed, or subject to legal and trade constraints—leave many communities untouched. Given these contingencies, where does the sense of pervasive technological change and, frequently, technological advancement in the face of new industrial objects or material properties come from?[1] In other words, what has historically counted as a new material in industrial cultures, to whom, and why?

The point of this collection, which gathers disparate cases of materials development from North America, Europe, the Middle East, and Asia over the last two hundred years and approaches the rubric "industrial" itself with criticality, is to see what social landscapes might emerge when we pay attention to the attention paid, historically, to "new materials." Both parts of that locution, "new" and "material," remain relatively undertheorized by historians of commerce and industry. This book is not a comprehensive survey of industrial innovation (however one might choose historically to delimit that category) nor is it meant to highlight materials that have had a singular impact on human experience (however one might choose to measure such an impact). Instead, the chapters in this volume suggest that when some matter is seen to be novel in a particular time and place, defined as a useful material for commerce, or simply delineated as the result of human artifice, a significant set of social enactments has occurred. Follow these enactments—whether the development of Quaker asylum locks, massive wood skeletons for high-rise buildings, or farmed catfish—and we find patterns of cultural change and accretions of power, both local and widespread. As well, a reflexive focus on the development of new orange breeds of the 1910s or carbon-fiber-reinforced polymer bicycle parts of the 2000s can help reveal our own analytical investments in particular distributions of goods or resources, as either writers or readers of material biographies who find some things and not others worth thinking about. A subset of recent scholarship focused on the "origins, careers, and misadventures of particular things," this collection asks how materials that emerge in societies that see themselves as modernizing—from Protestant medical missionaries' weatherproof clothing of 1915 to nanoscale drug ingredients a century later—have come to seem novel to particular audiences and what it means for historians to take up that actors' category.[2]

In the simplest sense, any perception of material novelty involves demarcating some matter or object as distinct from and perhaps preferable to some other. But for a sense of newness to prevail, that demarcation must also be foregrounded while attendant continuities in economic structures, political authority, or cultural influence are relegated to the background. In other words, "newness" is a *specific* way to demarcate a material (or, for that matter, a behavior or ideology) from others and one that serves its claimants by effacing that which is not to be renewed or replaced, that which is to be naturalized as ongoing. Tracking those continuities might easily be seen as the purview of academic history-writing and, in fact, the social and economic structures from which industrial invention has been seen to arise over time are familiar topics for historians of technology.³ Yet, the literature on instances of innovation and invention in industrialized societies has in large part taken shape without deeply problematizing *the cultural choice* to foreground some human activities and push others to the background. That is, only a small number of historians have questioned how it is that the very notion of novel materials comes to be deployed in a given historical moment. As David Edgerton helpfully summarizes: "The history of inventions we have is itself innovation-centric."[4]

In explaining how industry-derived materials have come into existence, much of this innovation-centric historical literature starts with the conditions of capitalist production as a given, tracing successful and sometimes also failed bids to control competitive markets or shape the natural landscape for profitable purposes. Other accounts focusing critically on the social origins and impacts of technological expansion have for some time shown such impulses to have costs as well as benefits, and that both are inequitably distributed; labor, race, gender, class relations, and imperial and colonial social structures are often well articulated through this literature.[5] Many of these contextualizing narratives chronicle complex and consequential episodes in industrial history and relatively few now fall back on older, teleological stories of human genius producing world-changing discovery. Yet, neither body of literature focuses on historically explaining actors' perpetuation of the category "new materials" (*as opposed to undertaking other materials-centered projects*) in political or institutional terms, leaving a significant explanatory gap.[6] Notable exceptions have, as Lissa Roberts and Schaffer put it, treated "the establishment and exploitation of explanatory categories" (here, the newness of materials or the desirability of change) as part of the history that needs to be told.[7] When historians cease to reproduce the effacements brought about by their actors' claims of newness, we can see such claims as enacting particular social and economic interests in the laboratories, drafting rooms, legislatures, committee meetings, board rooms, manufacturing plants, studios, workshops, greenhouses, and farms from which new materials emerge and in society at large. The manifold social instrumentalities of such projections of value, in thrall and in service to industrial, imperial, and other collective projects, can be explored.

New materials arising in industrial contexts, in other words, are not most fully investigated as expressions of modernizing impulses but rather as attempted distributions of merit, security, risks, resources, health, or mobility. Some features of this landscape of expectation are described in Jens Beckert's *Imagined Futures*, which captures the nature of capitalist economies as environments in which "current forms will not last" and firms, employees, and consumers must "be constantly oriented toward a future they cannot yet see."[8] The envisioning of preferred future conditions, both material and not, is an aspect of managing the future, and we can take from both historians of industrial capitalism such as Beckert and Cyrus Mody, and scholars of decolonialization such as Tuck and Yang, the idea that for those in authority and those with little social influence, materials and commitments to their development project some experiences at the expense of others.[9] The very broadest forces of development in modern industrial nations, including capitalism; consumerism; globalization; settlement; colonialism; imperialism; fascism; and raced, gendered, heteronormative and ablest ideologies are well understood by historians to have unjustly distributed the costs and benefits of mechanization and the dangers and fruits of science. The lifecycles of particular industrial objects, such as those described in this volume, elaborate some of those distributions. Notably, explicitly democratic and inclusive ideologies have also at times manifest themselves in the manufactured goods, infrastructures, and landscapes of industrial cultures with nonetheless decidedly mixed impacts on different communities.[10] The chapters that make up *New Materials* can thus be taken not just as a set of case studies of industrial interest in new materials but as examples of the ways in which the history of science and technology can serve questions of general history.

This volume is thus itself fully, and inevitably, a project of selectively sustaining the old and delineating the new. The authors included in *New Materials* are distinct in some ways from those who have previously analyzed mechanization or seen the creation of new technological systems in a critical light—long established as legitimate aims for scholars—in that they focus on the nature of matter in industrial cultures as such (we will see below that the term *raw material* does not serve us well here). They also offer energetic contributions to recent inquiries focused on materialities in sociology, anthropology, cultural studies, and science and technology studies (STS), demonstrating the explanatory richness that results from bringing history into that mix. For example, the idea that action and matter produce one another, growing from the work of Bruno Latour, Karen Barad, Jane Bennett, and others, has allowed STS to see modern constitutions of bodies, objects, and knowledge systems as deeply entangled with one another, and the authors in this collection show how such entanglements reflect the profound political contingency of industrial development: ontology over time.[11]

The foundational work of STS scholars including Latour, Michel Callon, John Law, and Madeleine Akrich, among others, emphasizing the multiple, indeterminate nature of thing/human encounters (and the very possibility of thingness, and humanness, in particular settings) is vital to this collection. That work has immeasurably helped the historical project of accounting for precise objects' "mode[s] of existence" and roles played "in the production of sociomaterial order," as Jerome Denis and David Pontille put it, and creates a vantage point from which materials may be seen to be producers (rather than expressions) of sociabilities.[12] Stasis and change are not in any sense outside the STS purview: the field's explications of technological systems and economic or geopolitical networks often center on accounting for material change, and its seemingly micro-scale inquiries (such as Denis and Pontille on rusting subway signage) are no less suggestive. Albena Yaneva, meanwhile, articulates the artifacts of technological or design enterprises as "nonsocial ties that are brought together to make the social durable," not least effectively through the encouragement of imitation and repetition over time.[13] But the aims of STS are not coterminous with those of history fields; as Tiago Saraiva has written, the so-called study of things, does not automatically involve "examining the forms within which these things operate in the world(s)."[14] This volume brings the causal, explanatory imperatives of history—attending to change and stasis on a wide and unpredictable array of geographic and time scales—together with the insistence, prompted by STS, that human subjects and their objects of attention are only arbitrarily separated in our narratives.

On the whole, the authors in *New Materials*, while writing historically, move beyond familiar historiographies of contextualized invention that have located durable cultural and economic interests driving industrial research and development. As Edgerton makes clear, such contextualism "assumes that the existing historical work used to establish context does not already have a particular account of technology in it."[15] The "particular account" embedded in many histories of new industrial materials is one of economic growth and indisputably legible technological change. By contrast, this collection takes up the question of why demarcations of industrial expansion and change, rather than problematics of, say, community security, state violence, global economic redistribution, universal design, or environmental sustainability, should direct narratives of invention.[16] Like the inventors, investors, farmers, missionaries, consumers, and workers described in this book, the authors and readers of these seven case studies invariably establish and exploit explanatory categories; we, too, are historical actors who reproduce existing social relations, or not. With the temporal curiosities of history in hand, episodes of material consistency and inconsistency offer what Alain Pottage has called a theme "of considerable critical potential," giving us "the resources to dissolve and recompose the premises of taken-for-granted categories that intervene before analysis gets underway."[17] Such dissolutions and recompositions are the shared aims of the seven chapters and afterword that follow.

THE VALUE(S) OF MATERIALS

The motivating questions behind this collection emerged from a workshop organized by the Hagley Museum and Library in 2011 on the history of new industrial materials; one of the authors included here was part of that event and I provided some commentary.[18] This book maintains one of that meeting's broad aims of studying industrial attention to materials through time, while also taking advantage of the reflective nature of written essays to problematize that rubric. In this way, we may see anew the distributions of economic, environmental, and cultural authority that inhere in directed industrial activity. Modern commitments to material productivity of course include not just those who control means of production but also those who produce, labor with, maintain, purchase, use, or derive pleasure and health from the things produced. Those who would devise means to compete with, evade, or destroy industrial technologies configure their histories as well. The division of encounters with industrial materials into "supply" and "demand" is deeply problematic also in obfuscating the experiences of those who have little choice but to interact with such artifacts in their home- or working lives; consider simply the materials used in water infrastructures, medical devices, and manufacturing processes. Thus, those who are potentially harmed by new materials can also in an important sense be instrumental in their emergence and proliferation (and profitability); we must account for bodies and minds put toward the making, deployment, or disposal of industrial products. This volume suggests some ways to account for such diverse collective engagements with material novelty over time. In thinking about this volume as itself a collectivity, it became impossible not to suggest that all of our authors are interested in how historical actors pursue "consistency" in both senses of that word: the identifying texture or material characteristics of the substances they study, such as heft, odor, or viscosity, and the intentional reproduction of such characteristics—what we might see as historical efforts to achieve conformity to type, along with the social circumstances that assure physical resemblance over time and space. There is a potent critical possibility in keeping both definitions in mind as our authors do, rendering physical and social reproduction as inseparable, synchronous intentionalities.[19]

Perhaps ironically, given the importance of objects and their physical characteristics to this project, the essays in this volume also follow actors' thinking about new materials and not just instances of new materials coming into being.[20] By attending to their actors' *wishes* for some things to exist, our authors open their inquiries to an array of historical conditions that have constituted "the new" or in which "new" has been an instrumental classification. Sociologists, anthropologists, and STS scholars have made clear that the social instrumentalities of newness include all of the following: supporting expedient claims to technical knowledge (as deployed in patent cases or medical diagnoses), formulations of market value (asserting worthwhile investment or consumption), and assertions of group identity.[21] Unlike most historians of technology who have recounted the emergence of novel things in modern industry, that scholarship has now made clear that the striving for consistency in no way precludes "multiple" or "mutating" objects and that even ostensibly standardized entities "are never unitary or stable."[22] The chapters in this collection take up those analytical possibilities and, with the priorities of history writing, also address the complex causal circumstances that bring about the yearning for and expectations of industrial novelty.

Some scholars in the history of technology, history of science, and business history have now, we should note, stepped away from market-oriented explanations of industrial innovation since 1800, and many do so to highlight the social aims driving the sorts of novelty that historically "preoccupy the privileged," as Andrew Russell and Lee Vinsel put it. For many Americans, we learn, these aims have historically included a nearly fetishistic attention to innovation that complements the veneration of private property and wide differentials in wealth.[23] Mody adds the compelling cultural mechanism "conditional prophecy," whereby inventors or investors project a scenario of demand for particular technological innovations going unmet, thus spurring their community's efforts (such as institutional innovation or the building of heterogeneous networks) at filling precisely that need. Again, supply and demand, even within corporate sectors, are revealed as actors' categories that we cannot adopt uncritically if we hope to problematize what does and does not draw the resources of capital and science.[24] The pieces in this volume, determined to expose the widest possible range of sociabilities involved in modern inventive efforts, couple those sorts of inquiry with established critical approaches drawn from social, political, and labor history; feminist and queer theory; and other areas of historical scholarship bent on helping us see who benefits and who suffers as technological ingenuity and social ideologies come to be. Twenty-first century developers' investments in unprecedentedly tall, and potentially unprecedentedly flammable, timber buildings; orange growers' labor practices; British missionaries' ideas of "indigenous" craft skills among colonial peoples; and European bicycle buyers' racialized ideas about Asian manufacturers are just some of the conceptions shaping material innovations in this book, all reasserting social hierarchies through impulses that are inseparable, our authors show, from what we might customarily call technical determinations.

For the authors in this volume, historically interesting instances of new industrial materials over the last two hundred years involve many kinds of matter: living organisms grown for human food; metallurgical innovations and known metals put to new purposes;

and cloth, wood, plastic, minerals, and flesh, as well as light, air, and water, all drawn into new configurations. Iron asylum locks, high-rise building-frame materials, nanopharmaceuticals, farmed fish, missionaries' clothing for cold climates, selectively bred orange plants, and carbon-based bicycle parts attain a certain symmetry in this collection and we are freed, at least for the moment, from sectoral analyses of invention centered on, say, the history of agricultural or construction or biomedical fields as isolated drivers of inventive efforts. Similarly, where processes of industrialization begin and end comes under scrutiny for our authors, with fruitful analytical results. In the case of historical labor relations, for example, we can ask: Considering that early-nineteenth-century blacksmiths working for Quaker clients faced customer expectations of standardized goods and practices, why precisely would we treat that artisanal labor as somehow *pre*industrial? What has that retrospective term come to mean in accounts of industrial lives? Values are expressed in all such framings by both actors and the historians who recount their experiences. With such generative heuristics, we are released from familiar delineations between bodies of empirical knowledge (say, between manufacturing, agriculture, and medicine) and stale binaries (preindustrial versus modern societies, human-made versus natural things), loosening our dependence on those "taken-for-granted categories."

In *New Materials*, we find a pronounced commitment to interrogating taxonomies of all kinds. The editors of the recent compendium *Objects and Materials: A Routledge Companion* have urged us to acknowledge "that any specific material form or entity with edges, surfaces or bounded integrity is not only provisional but also potentially transformative of other entities."[25] As I have written elsewhere, bringing this edge-attentive approach to labor history makes clear the nature of managerial authority: If we want to gauge what has historically constituted human capacity or responsibility *for* material action in a workplace (where such factors determine employment opportunities, conditions, wages, etc.) we must also determine where *materials'* capacities or responsibilities have been seen to end.[26] We will see that the problem of knowing where a "new material" begins and ends for historical actors is central to every case in this book, explicating social relations well beyond those directly concerned with labor. Scott Knowles and Jose Torero, focusing in this volume on risk and safety in commercial building sectors, note that blame for deadly fires in residential towers since the 1990s has rested variously on building materials, individuals, and regulatory organizations in every conceivable combination, with formulations of human and nonhuman behaviors in constant reconfiguration as forensic experts' explanatory aims might require. For Tiago Saraiva's orange growers of 1910s California, a new type of fruit comes into being, along with the damaging mold that afflicts it and the untrimmed fingernails of the laborers who pick it. In Karen Senaga's account of commercially produced catfish, the fish's optimal flavor is not a given but is instead the object of stabilizing efforts by producers eager to project consumer preferences. As Patryk Wasiak and Rafico Ruiz both show in their chapters, by the late nineteenth century, it becomes impossible for any internationally marketed object not to have a nationality, and nations are in turn knowable by the products they send out into the world, place and product mutually affirming and transforming, and so on through our cases.[27]

If ascriptions of skill or technical knowledge and distributions of productive responsibility become evident with such heightened attention to materials' "edges," so too do other prevailing definitions. For example, the idea that nature is that which is not made by humans, a notion thoroughly challenged by environmental historians and philosophers at this point, is further challenged by nearly every case in this book as the authors articulate shifting, ideologically freighted human-landscape or human-animal distinctions.[28] For instance, in their own view, Grenfell missionaries fought the cold of Labrador and Newfoundland in order to minister to deserving locals. But it is only by first demarcating oneself as an actor moving about within an ecosystem, as figure on ground, that such battles can be imagined. "Environmental mediation" cannot proceed without intentional delineation of particular human goals (in this case, functional missionaries) as projects stymied by nonhuman conditions. Across all of these attempts to stabilize edges—involving the boundaries of knowledge, matter, entities, and personhoods—actors' obsessions with the project of definitional fixity are clear. The mechanical operations of industrial production are literally repeatable, and so are material delineations and the social enactments of which they are part. But crucially, invention and innovation are not historically equivalent to all other kinds of social ordering. Those actions may achieve regularity in social conducts, but they must do so by appearing to imagine and produce some new thing. As noted above, this projection of the novel is a particular social instrumentality, and one that in industrial settings works especially well in the preservation of privilege for holders of capital, states, employers, and others in authority.[29]

In some sense, the collection's stance is experimental, challenging the value of "invention" and even "industrial material" as organizing concepts for histories of science, technology, or commerce; why not, our authors suggest, track distributions of risk, security, or economic mobility instead and treat novelty as a subordinate technique or effect of power? That sort of methodological question helps us avoid being unintentionally captured by the categories (such as "new," "material," or "new material") used by our historical actors. And yet we must still ask: How exactly has *technological innovation* come to displace other meaningful labels for thought and action in industrial cultures of the last two centuries? In choosing cases from the post-1800 era in which material innovations have supported the aims of industrialists and other influential figures in industry-centered cultures, this volume centers on a long period in which such innovations have been widely cast as sources of individual, institutional, and societal betterment.[30]

That obfuscating cultural equation of material change with societal well being and of increased economic value with personal welfare and virtue—along with the evident irony of such equations in light of a modern industrial history steeped transnationally in economic, health, and other inequities—are central elements in the chapters that follow.

SOME HISTORIOGRAPHIC PRIORITIES

The chapters offered in this volume undertake a number of historiographic projects in their revelation of global sociotechnological conditions since 1800. For example, rather than recapitulating their actors' rhetorical distinctions between raw materials and fabricated artifacts, the authors in this volume historicize demarcations between constituent elements and finished products. In one instance, the perceived availability of nanoscale gold particles (a raw material) makes the creation of a new pharmaceutical (a finished product) seem viable to developers. Meanwhile, in another case, prevailing ideas of tasty, profitably farmed foods on which to base the consumer's diet (finished products) render catfish (a raw material) a reasonable substance from which to mass-produce edible protein. Even within such delineations of raw and finished, the authors find indeterminacy: What shall actually count to stakeholders as definitive identifications of nanoparticles remains unsettled, as regulatory and market conditions face contestation. Clean-tasting scavenger fish are not a reasonable option to everyone to whom they are presented, meaning that, to some, even a package of "farmed catfish" held in the hand in the supermarket is not apprehendable as "dinner" but a conceptual impossibility, a mistake waiting to be made.[31]

For new materials to be recognized as such, all sorts of categorical work must be undertaken. Consider that it is only through the proposition of differing things being of the same category that "better" versions of things can be labeled as such.[32] This taxonomic activity is not a matter of abstraction from real physical conditions but a vital step in establishing narratives of success and failure. If the materials from which buildings' structural elements are fabricated (here, wood) are to be rendered culpable for high-rise fires, analysts must disarticulate the flawed structural elements from the economic or aesthetic priorities (say, environmentally sustainable construction or dramatic height) that brought those elements into existence. Presumably some other sorts of beams and columns, deriving from the same priorities, would not have caused fires or ones of such severity (of course other analysts might find the beams and columns innocent and blame the architects, fire department, or the putative tenant who smoked in bed).[33]

In recognizing the indeterminate nature of relationships between resources (whether material, institutional, or intellectual) and their deployments, as our authors do, we are liberated from "a prior polarity that would distribute reliable knowledge and effective action between two adamantine categories, roughly identifiable as science and technology," as Roberts and Schaffer articulate conventional historical discourse about European invention. That linear framing of the epistemics of invention, wherein basic knowledge precedes application and ideas precede action, hides the mutually formative character of knowledge and world, and, even more potently, of knowers and known-about.[34] For example, in accounting for materials research or production, the authors in *New Materials* identify insistent labor, trade, and regulatory systems as well as governmental regimes that are rarely invoked in any overt way by those actors immediately responsible for industrial operations. The most expert orange grower of 1920 California was he (always *he*) who most effectively minimized wage costs; but few such growers would have imagined that the aim of limiting labor expenditures merited discussion or debate. White communities of growers "naturally" enacted the economic subordination of Mexican workers and developed "improved" fruit types that suited such labor practices.

Yet these chapters do not treat social systems and regimes as background contexts for these inventive episodes but rather as conditions that are co-produced with materials and products.[35] Here, the essays in *New Materials* echo a handful of previous studies of industrial materials that have probed the social instrumentalities of attributes like the malleability of rubber and fluidity of concrete. Those characteristics would have had no meaning apart from imperial conceptions of commerce in nineteenth-century Europe (in which rubber played a large role) and the deskilling of building labor in early-twentieth-century United States (enacted through the development of commercial concrete), respectively. But nor would those regimes have played out without their material correlates.[36]

This historical collection offers, in short, a relational understanding of materials and human actors that is consonant with central questions of sociology, anthropology, and STS. Such inquiries are based on the idea that action does not occur against a fixed backdrop of materialities (whether that backdrop is conceptualized as nature, objects, stuff, or any other solidification of preexisting human interests). Rather, in John Law's phrasing, "Something becomes material because it makes a difference: because somehow or other it is detectable." In other words, material depends "on a relation between that which is detected and that which does the detecting."[37] The so-named activities of scientific investigation and characterization, the focus of Karen Senaga's and Sharon Ku's case studies that open the book, perhaps stand out as obvious sites of detection—detection ostensibly being, after all, the primary modality of scientific research and testing. As Andrew Pickering articulates, science, the "certified way of knowing" in Western industrial cultures, undertakes a veiling of the "symmetric ... process of the becoming of the human and non-human," a crucial part of a "dualist detachment" that hides its own social effects.[38] The authors in *New Materials* show us still more forms of this modern detection-as-veiling: the operations of mass production, as in the case of Saraiva's oranges or Knowles and Torero's structural building elements, have historically demarcated scientific subjects (say, materials experts) from scientific objects (including, among such objects, the materials studied by experts but also the nonexperts meant to deploy experts' findings). Further, the assertion of group identities, as in Darin Hayton's piece on Quaker asylum construction, Wasiak's on Taiwanese bicycle parts, or Ruiz's on the textile innovations of British medical missionaries, enacts the conjoined detection of materials and humans.[39]

In all of these episodes, the authors recognize what Barad labels as the "co-constitution of material and discursive constraints and exclusions," and, in so doing, they delineate minute social mechanisms that sustain wide flows of power and influence.[40] Using that foundational concept, an array of scholars in the humanities and social sciences now embrace the idea, as the editors of *Objects and Materials* put it, that "*things* are relational, that subject/object distinctions are produced through work of differentiation" (emphasis in original).[41] There is not, proponents of this orientation would say, a "single ordered cosmos" waiting to be discovered by scientists and engineers; instead, objects "come to be in relational, multiple, fluid, and more or less unordered and indeterminate (set of) specific and provisional practices."[42] Far from hiding or under-counting the impacts of human intentions, in allowing materialities to share the stage in this way with actors' particular curiosities and ambitions, we add tremendous specificity to our understanding of the social instrumentalities of research, experimentation, and testing, and of empiricism writ large in industrial societies. A relational approach to novel materials has great potential to reveal the social power that inheres in demarcations of some subjects from other subjects, and subjects from objects.

Nearly a century ago, Dewey's pragmatism, as Saraiva explains in his chapter, posited the democratizing possibilities of questioning subject/object dichotomies, and most of the authors here take up this reformist civic inclination.[43] Ontological leanings in STS support that project (Barad pinpoints the feminist genealogies of such "agential realism," for example), and I will touch on why this is so in the chapter overview, below.[44] But first, I want to consider one way in which such an indeterminate, relational approach to new industrial materials challenges familiar narratives of innovation and then touch on the moral implications of decentering invention in the history of technology.

If STS has lately offered us the possibility of seeing objects as possessing a kind of agency, a mode of "political analysis that can better account for the contributions of non-human actants," as Bennett puts it, what precisely happens to our understanding of invention and discovery as conceptual projects situated in particular communities and historical moments?[45] How does the importance of this category of human activity, long central to histories of technology and industry, change? Do figures such as Thomas Edison or George Washington Carver lose their exalted status? Do we begin to credit tungsten filaments and peanuts with ingenuity instead of the humans who we in the United States have previously celebrated as discovering the practical potential of such materials? And, do we risk denying the credibility correctively conferred by laudatory biography on those, like Carver, who have historically been marginalized?[46] Or, alternatively: Does the agency newly ascribed to neutrons and pulverized landscapes displace the agonizing responsibility Robert Oppenheimer ultimately felt for the development of the atomic bomb? These are clearly questions of considerable consequence for the moral lessons offered by historical scholarship.

It is not hard to see that a relational approach to industrial innovation has at least some moral utility because it helps us historicize our own valorization or condemnation of particular human inventive efforts. We render some things as the traces of valorous human ingenuity and others as the products of greed or cruelty; all of these judgments are inherently political and can be shown to result from our own historically situated analyses (i.e., we can recognize that we laud Edison's genius and not his filaments' eagerness to glow). But a still deeper critical possibility rapidly emerges because we are also empowered with this ontological approach to see human actions and material productions that have been subjected to *neither* praise nor condemnation. That is, we can recognize the deeply historical character of the detection Law refers to. For example, maintenance has long been arbitrarily omitted from accounts of the intellectual and physical labor of industry treated as seminal, as Edgerton and lately others like Russell and Vinsel have made clear.[47] Indeed, an entire shadow world of new materials, deeply consequential and steeped in the effects of power, becomes visible if we thoroughly historicize moments in which new materials have been seen as such. The sweat of orange pickers was not patented nor the sound of wind passing through commercial timber plantations; the effluents of nanogold production operations were not part of pharmaceutical marketing strategies; the snow that vexed the Grenfell missionaries on their ministering rounds was not subject to reengineering. Our authors launch for us the project of detecting these ghostly materials and materializations. Through a search for such phantom events and things, the precise structures of reward and oppression enacted in productive modern societies may become visible. Barad writes that "objectivity and agency are bound up with issues of responsibility and accountability," and such ghost hunts can perhaps be numbered among our own responsibilities.[48]

CHAPTER OVERVIEW

As the points above indicate, the chapters in this book can usefully be read in any order. Many themes and questions weave throughout; for example, all consider the simultaneous, inseparable actions of knowing and making worlds that constitute science, technology, and commerce. Each chapter helps us understand in historical terms how experts, owners, makers, purveyors, users, neighbors, or victims of new materials actively construct notions of what shall have meaning as a new material in a given time and place. As well, almost all the authors draw us into thinking deeply about the sensory, beyond the tactile experiences we might expect to figure in material studies to include taste, sound, smell, and, most intriguingly, tangled combinations of these experiences that add up to pain or pleasure as muscles, mouths, hands, and minds encounter new materials. But despite these connections across the chapters, and of the fearsome warning offered by one of our authors that "every time we divide we are lying," I have nonetheless organized the seven essays into three groups.[49]

Part 1, "Materials Tested, Success Defined," delineates the role of experimental and metrological practices in the production of new materials and the institutions that sustain them. The history of what is seen to *be* measurable figures at least as largely as materials researchers' specific findings in these chapters. Part 2, "Materials Produced, Labor Directed," centers on historical ascriptions of value to things and to the people responsible for their creation or deployment, whether as experts or subordinates in the workforce. In both chapters in part 2, a map of virtuous conducts for everything/everyone involved is projected alongside that production of market value and sustainable organizations. Finally, part 3, "Materials Interpreted, Communities Designed," draws our attention to the production of explicit resemblance among persons—that is, consistency across individuals' ascribed race, class, ethnicity, gender, nation, or ethical stance—through the circulation of new materials. Industrial products that are made to specifications and inspected assure predictable characteristics and durability, but so, too, religious groups and nations can be actively rendered identifiable and durable. Watching for each of these themes across all three sections will reward readers, but as divided here, those themes serve as windows to the social instrumentalities of new materials.

To begin with, the two chapters paired in part 1, "Materials

Tested/Success Defined," highlight the means by which industrial actors come to establish the character of a material in scientific terms—that is, to establish what shall count as metrics and evidence of its value, performance, or frailties. In the chapters offered by Karen Senaga and Sharon Ku, much of the action takes place in laboratories of the mid-twentieth to early-twenty-first centuries, sites dedicated to the development and refinement of one sort of definitive knowledge about materials. The two chapters on industrial materialities of farmed fish and nanoscale pharmacology, respectively, highlight what has historically counted as a credible setting for systematic inquiries into industrial materials, and, of course, who may credibly establish value for materials or knowledge and, not inconsequentially, bodies. But these two essays emphatically do not recount processes of mastery in any simple way: there is no apprehension by materials researchers or regulators of some singular "timeless essence" of any organism, of foods and the human mouths that taste them, or even of molecules and the drugs into which they combine.[50] Instead, we follow the production of tractable objects, as Ken Alder has succinctly described scientific research, by actors in pursuit of a kind of self-corroborating specialist understanding of materials.[51]

Pottage writes, "Material is sociality and sociality is always a process," and it is the incessant conversion of subjective experience into objective data that shapes the sociality of materials research in these two cases.[52] Senaga connects researchers' entanglements of materials, bodies, and quantification in her history of commercial catfish production in the American South after 1960. She makes clear that a mutually formative relation plays out in which cultural priorities highlight material conditions and material conditions render cultural priorities actionable. For catfish growers, perceived market demands also rendered competence and the knowing of material by certain individuals vital. Through enmeshed social and material quality controls, what was once a cheap foodstuff for poor and black communities might now reach more affluent and white markets, promoters believed. Senaga describes fraught encounters between scientists, growers, commercial processors, catfish, and the ponds in which catfish were farmed around the problem of "off-flavor": a supposed muddy or fishy taste associated with catfish found in the wild and identified with the gustatory preferences of the poor who caught and ate them. The catfish were resistant to dietary instruction: growers bent on using only commercial fish foods had difficulty controlling what the scavenger (and cannibalistic) species actually consumed in the crowded, outdoor ponds. Writing on the large-scale commercial rearing of swine in crates, Dawn Coppin has raised the issue of where animals' agency resides in the world of farming; she discourages us from seeing agency as "an attribute that an entity either inherently possesses or does not," because "it is not possible to subtract one thing [swine, catfish, human] that will retain its agency outside the relationship."[53] Senaga's case confirms for us that although it essentialized and instrumentalized animals, researchers' approaches to assuring marketable products cannot be understood without attention to the agency of recalcitrant fish.

Senaga shares with us stories of attempts to render unruly "nature" (plants, chemicals, bodily functions) subject to the plans of technical experts and other authoritative actors—a theme we will see playing out in most of our later chapters as well. Not a few of these attempts collapse in the face of social and material obstacles, with laboratories, corporations, and investment schemes stymied. Ku's chapter on the development of private and federal standards for nanogold particle measurement, critical for biomedical applications over the last decade or so, digs deeply into one such near- or yet-to-happen collapse. She follows the recent history of nanoscale particles and their measurement as a series of bureaucratic projects on the parts of medical researchers and global standardizing organizations. As Ku clarifies, measurement and other attempts to regularize materials of use to industry (efforts that, not incidentally, can precede the investigation and even the identification of material subjects) are acts that readily amplify and multiply disagreements. Ku adds the creation of techno-bureaucracies to the labor undertaken by researchers at benchtops and their sponsors in the boardroom or legislature. But if these organizational structures (from paper documents to federal agencies) are at times brought into existence in order to support the resolution of such disagreements, they are nonetheless fragile infrastructures that require constant maintenance.

There is also, Ku points out, a way in which data has an analgesic function. Its existence or the hope of its existence reassures that order can be reasonably imagined. Acts of systematic measuring and any resultant knowing feel good and are good, both for those who undertake them and for larger collectivities. In the two chapters of part 2, "Materials Produced, Labor Directed," Tiago Saraiva, Scott Knowles, and Jose Torero help us follow the long twentieth century's production of standardized, large-scale commodities; in both chapters, inseparable efforts by experts to order materials *and* organizations, rather than reliance on abstracted economic ideologies, are found to center efforts at technological innovation. Ascriptions of blame for material failures, damaged goods, and other reversals of commercial fortune offer a superb lens on the social relations of technical knowledge-making. Crucially, in both Saraiva's study of California orange-grower cooperatives and in Knowles and Torero's chapter on the emergent fashion for timber skyscrapers, multiple collectivities make up a given social landscape. Aims of civic well-being—a kind of social stewardship ironically associated for many with the individualist impulses of US capitalism—nominally drive all of the expert groups described, working nearly a century apart in the two cases. Yet, perhaps unsurprisingly, sharp divergences of what shall constitute social good are elaborated, even within each account.

As described by Saraiva, California scientists who developed new orange breeds starting in the 1910s and growers who took up those new crops integrated farming for the market with a far wider social project: the instantiation of US-style democratic ideologies. These would find expression in the application of state-sponsored bioscience to crop improvement, in the creation of cooperative organizations and communities for white orange producers, and the production of idealized images of standardized homes and family structures for the predominantly Mexican pool of orange pickers. That perceptions of racial difference divided these communities, relegating those of "non-white" heritage to the low-paid and insecure work of the fields, was taken by planners and growers as natural and not contradictory to the tenets of industrial democracy. With his stress on the orange as "new material," Saraiva reveals minute and routine enactments of these racial divides: ideas of where orange flesh ended and the picker's hand began were embodied in scientific studies of the fruit's vulnerabilities to "mishandling," and field practices enacted profoundly inequitable labor relations around the resultant findings. We find that achieving quality control of the oranges through the science of breeding and growing brought consistency as well to farmworkers, or at least lent power to growers' ideals of such connected racial and behavioral uniformity. Valuable orange harvests wrought virtuous people of both owners and workers and vice versa. This was for white scientists and growers an unassailable vision of democratic industry and, we should specify, industriousness.[54]

Knowles and Torero take up other episodes of (attempted) industrial standardization: the production and deployment of systems for building high-rises out of cross-laminated wood products. Emerging decades after slow-burning wood construction lost favor to the use of concrete and steel for large-scale structures, this new-old material is poised to replace the use of materials—specifically, mass-fabricated reinforced concrete and steel building elements—that have long stood at the forefront of modernist high-rise architecture. The new timber products are widely believed to offer environmentally sustainable alternatives, their aura green and "wholesome," as the authors explain. And yet, in a sort of return of the repressed, ancient fears about the fire risks of wooden construction are not easily displaced for insurers, fire departments, or building occupants to whom architects and developers pitch the timber towers. Knowles and Torero track historical distributions of responsibility and blame for such disasters, both real and potential, among different groups of people and categories of things in different disaster scenarios. A truly complex ethical landscape emerges for readers: novel construction materials and techniques today emerge amid assumptions of competency that sustain notably *risk-tolerant* building and regulatory operations in growing global cities. Ironically, the pro-wood values of environmental sustainability find fertile ground where other ethical priorities, such as safety, may be in recession.

If Saraiva's case allows us to see the manifold, and contradictory, features of white-led democratic reform in the progressive-era United States, Knowles and Torero follow deeply contradictory value systems driving urban building in the twenty-first century and the variable social instrumentalities of novelty in the hands of architects, engineers, and those responsible for building safety. Explicit in both chapters is the inescapable linkage of technical practices centered on material innovation, and the planning of organizational forms. But linkage does not equal stability: not only do ideas of technological innovation and security vary *across* settings, but pressures exerted by markets and social networks, understandings of risk, and competing self-estimations of occupational importance constantly combine and recombine *within* any given setting. The influence levels of particular communities and organizations ebb and flow. In this landscape, new materials come into being to celebrate cutting-edge technical and aesthetic priorities, but also to replace, to *displace*, those seen as having failed or as being likely to fail in the future. Hope and dread together drive these worlds of, as Saraiva puts it, "entangled social and scientific experimentalism."

The final three chapters in this collection turn directly to the idea that the edges of materials and recognizable social identities are arbitrarily drawn. The three historical cases in "Materials Interpreted/Communities Designed" span two centuries and several hemispheres, but each makes clear how a commitment to material innovation articulates both what and whom shall have social significance in a particular cultural episode. Each chapter explores the replacement of a familiar material with a new one for a known use—for constructing asylum buildings, for high-end bicycle parts, or for weatherproof clothing—in a way that also establishes group identity. This makes for a more complex understanding of identity politics (of either the privileged or oppressed) than we usually encounter in histories of difference, as materialities establish and carry forward specific attributes of their developers.

Darin Hayton, for example, describes building materials developed by Quaker asylum designers striving to improve patients' experience of mental illness and incarceration in early-nineteenth-century Pennsylvania. The Quakers actively defined themselves not only as uniquely compassionate caretakers of the unwell, but as members of a healthy, virtuous, and knowledgeable group within a polity made up of other social groups manifesting other traits. The thoughtful caring that constituted Quakerness in 1815 found expression in material choices for the asylum building, which involved defining attributes of both the caregivers and the objects of their attention—that is, attributes of both humans and materials. The materials used for walls, window sashes, door locks, and miscellaneous fasteners could all exert punitive conditions on patients or help restore their health. But consider the implications, as Hayton does, of the asylum sponsors' "Sub-Committee on Light and Air," which bluntly and drastically expands the material purview of architects beyond bricks and mortar to include ambience itself. Clearly, none of the architectural elements under study by the Quakers existed apart from the worlds they mediated; this was heating, ventilation, and air-conditioning (HVAC) by a more self-aware—and more lyrical—name, perhaps.

Boundaries between the hospital building's inside and outside are obviously at issue here, with doors and windows providing apertures vital for patient health and ease and, at the same time, the potential for self-harm or escape. If, to the subcommittee, a new type of metal window sash that looked like a conventional wooden one provided greater safety to the ill, it also imparted a more kindly and restorative aura to the facility than would metal bars. Hayton demonstrates that with such determinations, Quakers also produced themselves as exceptionally kind and adept healers of the mentally ill. We know metal by its hardness and sturdiness, but its history as a material in relation to humans reveals it to have produced a durable Quaker identity as well.

Of course, others had to corroborate both the meaningfulness of the category "Quaker" and the architectural choices made for the asylum. Patients', patrons', and neighbors' reactions to the asylum building were consequential for the demarcation of the Quakers as a group. They were also consequential for the self-demarcation of other groups; the materials of building and manufacturing enact identities beyond that of their producers. For Patryk Wasiak, the global flow of commodities at the turn of the twenty-first century—here, high-tech bicycle parts made of carbon-fiber-reinforced polymer by Taiwanese firms—produced a great many identities: the makers' own sense of themselves as a community of skilled metallurgists and fabricators; European bicyclists' sense of the Taiwanese as adept or, if unsatisfied with their purchases, inept artisans; and the consumers' sense of their own athletic prowess, rights as consumers, and non–Taiwanese nationality. With that array of community-bounding projects, we cannot be surprised that the phrase "Made in Taiwan" has had a radically shifting implication in the history of bicycle part manufacture. But Wasiak also shows that there is no material that is free of (or unempowered by?) race, nation, or other categories.

In the final chapter of this volume, Rafico Ruiz follows the movement of Grenfell Cloth, a sturdy, weatherproof textile devised by a British medical missionary working after 1900 in harsh climates of Newfoundland and Labrador, across many users, including soldiers during World War One, explorers on Everest, modern-day consumers, and, crucially, the historian working today in the Grenfell archives. Grenfell Cloth has upheld a notable ability over generations to "press its messages on attentive auditors," as Lorraine Daston has phrased the ability of things to carry nearly indisputable implications.[55] Ruiz's attention to transmedia manifestations of the cloth shows how, in each time and place, the fabric (scientifically derived and then woven on special looms) has constructed meritorious audiences for its commendable characteristics. To date, these have included the early-twentieth-century wearer adhering, with the textile's help, to standards for sturdy, geographically mobile Christian male bodies; later affluent buyers of Grenfell leisure wear who achieved the means and time for such discretionary adventure; the cloth's manufacturers with their commitments to family and heritage expressed through their internet "presence"; and today's scholar tracing those reproductions and rendering Grenfell Cloth worthy of our contemporary attention. Again, as in other chapters in the book, value and virtue are co-produced, identity and material reasserting each other's (shifting but incessant) significance.

Ruiz shows that Grenfell Cloth emerged and thrived through coordinated judgments about communities and matter. For example, the British missionaries established many employment opportunities for fisherfolk, women, and other subordinates-among-the-subordinate in Labrador, and the raw or recycled materials of such labor—some local, some donated from abroad—established both the workers' and the British philanthropists' virtue, but, obviously, no equivalence between the two. Communities are defined relentlessly through the circulation of Grenfell Cloth and narratives about it: the contingent transmedia life of the textile continues today in its makers' family website, in ascriptions of the cloth's "Britishness" in a recent branding campaign, and, not least, by the inclusion of Ruiz's text in this historical collection as a critical, geopolitically concerned account worth (say its editor and publishers, at least) reading.

Our own stories about, our own concerns with, new materials, in other words, are never outside the play of historical forces. For example, we also return in this final case to explicit engagements with "experimentation," best understood in Ruiz's telling *not* as investigations of nature but as the production of objects of study through systematic empirical attention: wind and snow (empirically produced as cold and threatening); human bodies of nonindigenous background that had to move through the Labrador landscape without traditional garments (righteous, willing, physically vulnerable); and new fabrics that might mediate between weather and missionary (much needed). As do the other investigations of materials and applications recounted in this book, the historiography of Grenfell Cloth includes our attention to that substance-making.

We might, then, understand this book as itself a project of making something new, enhancing through the writing and reading of history the value of our own interpretive efforts (and thus producing legitimacy for our salaries, credentials, and future working selves), along with significance for the materials to which we pay attention. But in addition to the criticality and reflection that the recognition of novelty may induce, possibly we can also bring about some dissolution and recomposition so that the new does not efface the old. Many of the activities that historically go into making new materials, as I have tried to stress in this introduction, represent the continuation of social structures and distributions of power—a consistency that is often deeply inequitable in its origins and effects. How can we make sure that the case studies that follow help us challenge that stasis? Projit Mukharji, author of our afterword, has written on the cultural specificity of novelty and indeed chronology as literary techniques for retrospectively organizing events through narrative. Mukharji reminds us that "tangible" and "intangible" are not conditions attained (either permanently or temporarily) by the objects of historians' attention but are instead the ascriptions of interested actors.[56] With such reminders, we may be more inclined to spot the ends to which claims of tangibility are directed, and the power and privilege that have accrued to those producing what come to be known as new materials. These final disturbances to our linear narratives of industrial innovation may help us with such a challenge.

In his brief 2007 essay, "Can We Get Our Materialities Back, Please?" Latour speaks to the mystification accomplished by the things humans produce: "Parts hide one another; and when the artifact is completed the activity that fit them together disappears entirely."[57] The chapters in *New Materials* assemble parts into narratives and arguments but often insist that we pay attention to the authors' activity that produced those final products: their choice to attend to some materials and not others, to some features of their chosen materials and not others, to some actors and events and not others. If readers find suggestive new concepts in this collection, as I hope they will, I also hope they will ask about what does and does not change for them with those innovative ideas in hand, critically reflecting on their activity along with our authors.

NOTES

1. David Edgerton, *The Shock of the Old* (Oxford: Oxford University Press, 2007), x; Jennifer Alexander, *The Mantra of Efficiency: From Water Wheel to Social Control* (Baltimore, MD: Johns Hopkins University Press, 2008), 1–14.
2. Ken Alder, "Focus: Thick Things, Introduction," *Isis* 98 (2007): 81; see also Lorraine Daston, ed., *Things That Talk: Object Lessons from Art and Science* (Brooklyn: Zone, 2004).
3. For helpful overviews of this vast literature, see Edgerton, *Shock of the Old* and Sally H. Clarke, Naomi R. Lamoreaux, and Steven W. Usselman, eds., *The Challenge of Remaining Innovative* (Palo Alto, CA: Stanford University Press, 2009).

4. Edgerton, *Shock of the Old*, 184; David Edgerton, "Time, Money and History," *Isis* 103, no. 2 (2012): 316–27. On the circulation of promising rhetoric about future technologies among western cultural arbiters, see Patrick McCray, *The Visioneers* (Princeton, NJ: Princeton University Press, 2012).
5. For overviews and introductions, see Donald Mackenzie and Judy Wajcman, eds., *The Social Shaping of Technology*, 2nd ed. (Buckingham and Philadelphia, PA: Open University Press, 1999); Harry Collins and Trevor Pinch, *The Golem at Large: What You Should Know about Technology* (Cambridge: Cambridge University Press, 1998); and Gabrielle Hecht, ed., *Entangled Geographies* (Cambridge, MA: MIT Press). Robert Friedel's *A Culture of Improvement: Technology and the Western Millennium* (Cambridge, MA: MIT Press, 2007) demonstrates the importance of attending to interim stages in technological development, rather than only to final products, if a sequence of technological innovations is to be established.
6. Tiago Saraiva, "The History of Cybernetics in McOndo," *History and Technology* 28, no. 4 (2012): 413–20.
7. Lissa Roberts and Simon Schaffer, preface to *The Mindful Hand: Inquiry and Invention from the Late Renaissance to Early Industrialization*, eds. Lissa Roberts, Simon Schaffer, and Peter Dear (Amsterdam: Royal Academy of Arts and Sciences, 2007), xv. Models of this kind of historical scholarship include Simon Schaffer, "'The Charter'd Thames': Naval Architecture and Experimental Spaces in Georgian Britain," in *Mindful Hand*, 279–305; Simon Schaffer, "Accurate Measurement is an English Science," in *The Values of Precision*, ed. M. Norton Wise (Princeton, NJ: Princeton University Press, 1995), 135–72; Ken Alder, *Engineering the Revolution: Arms and Enlightenment in France, 1763–1815* (Princeton, NJ: Princeton University Press, 1997); Tiago Saraiva, *Fascist Pigs: Techno-Scientific Organisms and the History of Fascism* (Cambridge, MA: MIT Press, 2016); Cyrus Mody, *The Long Arm of Moore's Law* (Cambridge, MA: MIT Press, forthcoming); and Chandra Mukerji, *Impossible Engineering: Technology and Territoriality on the Canal du Midi* (Princeton, NJ: Princeton University Press, 2009).
8. Jens Beckert, *Imagined Futures: Fictional Expectations and Capitalist Dynamics* (Cambridge, MA: Harvard University Press, 2016), 5.
9. Beckert, *Imagined Futures*; Mody, *Long Arm*. Also, on the role of imagined technological futures, see Simone M. Müller and Heidi J. S. Tworek, "Imagined Use as a Category of Analysis: New Approaches to the History of Technology," *History and Technology* 2, no. 2 (2016): 105–19. On moves to "alleviate the impacts of colonization" that in fact constitute attempts "to rescue settler futurity," see Eve Tuck and K. Wayne Yang, "Decolonialization Is Not a

Metaphor," *Decolonialization: Indigeneity, Education and Society* 1, no. 1 (2012): 3.

10. Ann Laura Stoler, ed., *Haunted by Empire* (Durham, NC: Duke University Press, 2006); Leigh Patel, "Reaching Beyond Democracy in Educational Policy Analysis," *Educational Policy* 30, no. 1 (2016): 114–27; Anne McClintock, *Imperial Leather: Race, Gender and Sexuality in the Colonial Context* (London and New York: Routledge, 1995).

11. Bruno Latour, *Reassembling the Social: An Introduction to Actor-Network Theory* (Oxford: Oxford University Press, 2005); Karen Barad, "Agential Realism: Feminist Interventions in Understanding Scientific Practices," in *The Science Studies Reader*, ed. Mario Biagiolli (London and New York: Routledge, 1999), 1–11; Jane Bennett, *Vibrant Matter: A Political Ecology of Things* (Durham, NC, and London: Duke University Press, 2009); Yvonne Marshall and Benjamin Alberti, "A Matter of Difference: Karen Barad, Ontology, and Archeological Bodies," *Cambridge Archeological Journal* 24, no. 1 (2014): 19–36; Bill Brown, *Things* (Chicago, IL: University of Chicago Press, 2004); Trevor Pinch, "Technology and Institutions: Living in a Material World," *Theory and Society* 37, no. 5 (2008): 461–83.

12. Latour, *Reassembling the Social*; Madeleine Akrich and Bruno Latour, "A Summary of a Convenient Vocabulary for the Semiotics of Human and Non-Human Assemblies," in *Shaping Technology/Building Society: Studies in Sociotechnical Change*, eds. Wiebe Bijker and John Law (Cambridge, MA: MIT Press, 1994): 259–64; Jerome Denis and David Pontille, "Material Ordering and the Care of Things," *Science, Technology and Human Values* 40, no. 3 (2015): 2.

13. Denis and Pontille, "Material Ordering"; John Law and Michel Callon, "The Life and Death of An Aircraft: A Network Analysis of Technical Change," in *Shaping Technology*, 21–52; Albena Yaneva, "Making the Social Hold: Towards an Actor-Network Theory of Design," *Design and Culture* 1, no. 3 (2009): 28.

14. Saraiva, *Fascist Pigs*, 258n80.

15. David Edgerton, "Innovation, Technology, or History: What is the Historiography of Technology About?" *Technology and Culture* 51 (2010): 694.

16. Mody, *Long Arm*; Aimi Hamraie, *Building Access: Universal Design and the Politics of Disability* (Minneapolis: University of Minnesota Press, 2017); David Edgerton, "The 'Linear Model' Did Not Exist: Reflections on the History and Historiography of Science and Research in Industry in the Twentieth Century," in *The Science-Industry Nexus: History, Policy, Implications*, eds. Karl Grandin, Nina Wormbs, and Sven Widmalm (Sagamore Beach, MA: Science History Publications, 2004), 1–36.

17. Alain Pottage, "The Materiality of What?" *Journal of Law and Society* 39, no. 1 (2012): 169.

18. At the 2011 meeting, Patryk Wasiak delivered versions of the papers in this volume; I participated as a commentator.
19. Roberts and Schaffer, preface, xv; Lorraine Daston, "Introduction: Speechless," in *Things That Talk*, 7–24; Stefan Timmermans and Steven Epstein, "A World of Standards but Not a Standard World: Toward a Sociology of Standards and Standardization," *Annual Review of Sociology* 36, no. 1 (2010): 69–89; Ian Hodder, "The Entanglements of Humans and Things: A Long-Term View," *New Literary History* 45, no. 1 (2014): 19–36; Amy E. Slaton, "Style/Type/Standard: The Production of Technological Resemblance," in *Picturing Science, Producing Art*, eds. Caroline A. Jones and Peter Galison (New York and London: Routledge, 1998), 78–97. In discussing how material figures in ritual, which in turn stabilizes social relations, Grant Kien writes, "We cannot live the same moment twice, but we can act out the same fictionalized moment ad infinitum" ("Actor-Network Theory: Translation as Material Culture," in *Material Cultural and Technology in Everyday Life: Ethnographic Approaches*, ed. Philip Vannini [New York: Peter Lang, 2009], 27).
20. Alexander, *Mantra of Efficiency*, 12.
21. Beckert, *Imagined Futures*; Geoffrey Bowker, "What's in a Patent?" in *Shaping Technology*, 58–59; Stefan Timmermans and Marc Berg, "Standardization in Action: Achieving Local Universality through Medical Protocols," *Social Studies of Science* 27, no. 2 (1997): 273–305; Annemarie Mol, *The Body Multiple: Ontology in Medical Practice* (Durham, NC, and London: Duke University Press, 2002); Donald MacKenzie, *Mechanizing Proof: Computing, Risk and Trust* (Cambridge, MA: MIT Press, 2001); Elizabeth B. Silva, "Haunting in the Material of Everyday Life," in *Objects and Materials: A Routledge Companion*, eds. Penny Harvey at al. (London and New York: Routledge, 2014), 187–96; Pinch, "Technology and Institutions."
22. Penny Harvey and Hannah Knox, "Objects and Materials: An Introduction," in *Objects and Materials*, 9.
23. Andrew Russell and Lee Vinsel, "Hail the Maintainers," Aeon, April 7, 2016, accessed August 18, 2016, https://aeon.co/essays/innovation-is-overvalued-maintenance-often-matters-more; Edgerton, *Shock of the Old*, is central to this critical historiography.
24. Mody, *Long Arm*, 25–26.
25. Harvey and Knox, "Objects and Materials," 1.
26. Amy E. Slaton, *All Good People: Diversity, Difference and Opportunity in High-Tech America* (Cambridge, MA: MIT Press, in preparation).
27. Suggestively, Marshall and Alberti problematize archeologists' customary boundaries between "bodies and clothing, nature and culture" but

"nonetheless [locate] difference in the subject, arguing for the existence of multiple subject positions" ("A Matter of Difference," 20). Similarly, on the arbitrary delineation of hair and clothing as aspects of legible bodies, see Rebecca Herzig, *Plucked: A History of Hair Removal* (New York: New York University Press, 2015).

28. William Cronon's *Nature's Metropolis: Chicago and the Great West* powerfully established the arbitrary material ascriptions involved in commodifying grain, lumber, and other "natural" entities (New York: W.W. Norton, 1991); Donna Haraway, *When Species Meet* (Minneapolis and London: University of Minnesota Press, 2007); Saraiva, *Fascist Pigs*.
29. See note 7, above; Edgerton; "Time, Money, and History"; Beckert, *Imagined Futures*. Feminist historians of technology have shown this to be the case with seemingly mundane materials of hygiene and domestic maintenance; Judith Wajcman, *Feminism Confronts Technology* (University Park: Pennsylvania State College Press, 1991); Susan Strasser, *Waste and Want: A Social History of Trash* (New York: Metropolitan Books, 1999); Herzig, *Plucked*.
30. Friedel, *Culture of Improvement*.
31. Harvey and Knox, "Objects and Materials," 4.
32. Geoffrey C. Bowker and Susan Leigh Star, *Sorting Things Out: Classification and Its Consequences* (Cambridge, MA: MIT Press, 1999); Amy E. Slaton, *Reinforced Concrete and the Modernization of American Building, 1900–1930* (Baltimore, MD: Johns Hopkins University Press, 2001).
33. I thank Scott Knowles for sharing this taxonomy of blame.
34. Roberts and Schaffer, preface, xix; Andrew Pickering, "New Ontologies," in *The Mangle in Practice: Science, Society, and Becoming*, eds. Andrew Pickering and Keith Guzik (Durham, NC, and London: Duke University Press, 2008), 8.
35. Willem Schinkel, "Sociological Discourse of the Relational: The Cases of Bourdieu and Latour," *The Sociological Review* 55, no. 4 (2007): 707–29.
36. Londa Schiebinger, *Plants and Empire: Colonial Bioprospecting in the Atlantic World* (Cambridge, MA: Harvard University Press, 2004); Slaton, *Reinforced Concrete*.
37. John Law, "The Materials of STS," *Heterogeneities*, April 9, 2009, pg. 1, accessed on August 16, 2016, http://www.heterogeneities.net/publications/Law2008MaterialsofSTS.pdf.
38. Pickering, "New Ontologies," 8.
39. Slaton, *All Good People*.
40. Barad, "Agential Realism," 6.
41. Harvey and Knox, "Objects and Materials," 1.
42. John Law and Mary Elisabeth Lien, "Slippery: Field Notes in Empirical

Ontology," *Social Studies of Science* 43 (2013): 365; for this approach as unusually embedded in historical analysis, see Simon Schaffer, "A Science Whose Business Is Bursting: Soap Bubbles as Commodities in Classical Physics," in *Things That Talk*, 147–92.

43. Mukharji, in "Occulted Materialities," 34, explores facticity itself a profoundly consequential epistemic commitment that conforms current STS.
44. Barad, "Agential Realism," 7.
45. Bennett, *Vibrant Matter*, x.
46. The functions of such laudatory biography include the reproduction of social marginalities along lines of race, gender, sexuality, and other lines; see Amy E. Slaton and Evelynn Hammonds, "From 'Missing Persons' to 'Critical Biography': Race and Gender in the History of Science, Technology and Medicine" (unpublished manuscript, 2016); Amy E. Slaton, *Race, Rigor and Selectivity: The History of an Occupational Color Line* (Cambridge, MA: Harvard University Press, 2010).
47. Edgerton, *Shock of the Old*, 75–102; Russell and Vinsel, "Hail the Maintainers."
48. Barad, "Agential Realism," 7; Avery Gordon, *Ghostly Matters: Haunting and the Sociological Imagination* (Minneapolis: University of Minnesota Press, 1997); Silva, "Haunting in the Material."
49. Tiago Saraiva, conversation with the author, July 28, 2016.
50. Pickering, "New Ontologies," 7.
51. Alder, "Focus," 82.
52. Alain Pottage, "Scale Models, Forensic Materiality and the Making of Modern Patent Law," *Social Studies of Science* 41, no. 5 (2011): 636–37.
53. Dawn Coppin, "Crate and Mangle: Questions of Agency in Confinement Livestock Facilities," in *Mangle in Practice*, 48–49.
54. The association of taste (as in *good taste* or cultural capital) and taste (as the bodily apprehension of some specific material character) are laid out in Pierre Bourdieu, *Distinction: A Social Critique of the Judgement of Taste* (Cambridge, MA: Harvard University Press, 1996 [1979]). The impossibility of disaggregating individuals' cultural competence and material products in industrial contexts is explored in Slaton, *All Good People*, through the idea of "capacity."
55. Lorraine Daston, "Introduction," *Things That Talk*, 12–13.
56. Projit Bihari Mukharji, "Occulted Materialities," *History and Technology* 34 (2018): 31–40.
57. Bruno Latour, "Focus: Thick Things, Can We Get Our Materialities Back, Please?" *Isis* 98 (2007): 98.

PART I

MATERIALS TESTED, SUCCESS DEFINED

CHAPTER TWO

MUDDY TO CLEAN

The Farm-Raised Catfish Industry, Agricultural Science, and Food Technologies

Karen Senaga

On a fall day in 1968, Joe Glover and Chester "Check" Stephens smelled something weird as the two men loaded catfish from a farmer's pond into water tanks in the back of their truck.[1] Stephens turned to Glover and asked, "Say, Joe, what do you suppose that smell is?" They did not know what it was, but they continued to work. Glover and Stephens drove the crops from the farm near Selma, Alabama, to their cramped processing facilities in Greensboro, where workers skinned, dismembered, and froze the fish. The two men, along with Richard True, owned one of the nation's first pond-raised catfish processing plants, which they named STRAL, a composite of their last names.[2] STRAL's moderate success had caught the attention of businesses like the Quaker Oats Company that wanted to get into catfish production too. On the very day that Glover and Stephens returned from their errand to the foul-smelling farm, a Quaker Oats representative was at the plant to take a sample of STRAL catfish. That night, Stephens received an alarming phone call: "They're the smelliest fish in the world! I just cooked some up for dinner, and my whole house smells like it's been fumigated! We can't eat them!" the Quaker Oats man on the phone shouted to Stephens. Stephens drove to the processing plant, picked up a box of fish from the same Selma batch, and cooked it up. The Quaker Oats rep was right: the fish, to Stephens, tasted terrible.[3]

The owners of STRAL learned a lesson and, henceforth, they would only take fish crops from ponds that they had sampled first. But their decision caused turmoil among suppliers: some catfish farmers simply did not believe that their fish tasted bad. When STRAL rejected what they found to be awful catfish, they had to fight angry farmers. "What do you mean?" one farmer furiously demanded in late 1968. "There's nothing wrong with these fish. They're good! Why, we've eaten them ourselves!" With some persuading, the farmer convinced Stephens to come back and taste the fish after a week or so. After a week, Stephens concluded that the catfish still tasted objectionable. Convinced that his own palate was just as good a judge for tasty or displeasing flavors, the farmer asserted, "Now, that's good fish. Nothing wrong with those fish!" Stephens snubbed the farmer's sensibilities and his ability to grow good fish. Finally, after a few weeks and heavy rains, Stephens tested the farmer's fish crops again. That time, the "musty flavor was gone."[4]

For those bent on mass-production and marketing, the catfish's flavors were elusive, subjective, ephemeral, and, at times, idiosyncratic. Glover and Stephens's interactions with the Quaker Oats representative and the disgruntled catfish farmer demonstrated how these off-flavors could potentially hinder industry growth.[5] From the 1960s to the present, farmers', scientists', and professional taste testers' senses were the primary tools in the standardization and categorization of the constructed and contested farm-raised catfish flavor. These stakeholders studied the causes of off-flavors, tried to develop technologies to inhibit displeasing flavors in the crop, and attempted to cultivate mechanisms to guarantee a consistent agricultural product. Researchers, farmers, and processors tried to impress their ideal catfish taste and smell upon the fish's flesh but disagreed on the optimal taste and smell. This ideal was akin to a subjective mild blandness and a nonfishy flesh that tasted more like land-based and grain-fed livestock such as chicken, rather than seafood. All considered the perception of what they characterized as "muddy" flavor as bad, although intensity was often debated. But the pursuit of their ideals proved even more onerous as the material reality of the fish and the waters consistently fought back.[6] No farmer controlled the animal in nature, but as researchers realized, without guidance and intervention, growers could barely control the fish within the pond environment. Despite the occasional strongly flavored piece of meat let loose into the market and onto consumer plates, the industry ultimately succeeded in ensuring that their brand of a bland, nonfishy meat entered consumers' mouths.

A study of the industry angling for the perfect farm-raised flavor uncovers how scientists', processors', and farmers' bodies and sensory experiences, all of which were loaded with cultural understandings of what was good and bad, were the "tools and agents" against off-flavors.[7] This study demonstrates how farmed catfish as a "new" material embodied researchers', processors', and farmers' own beliefs in differences between social groups. Their conflicted journey for a specific farmed flavor embodied classed and racialized understandings of what was deemed acceptable and unacceptable farmed flavor. The flavors that these various stakeholders encountered and judged were produced by the environmental contingencies of the pond and the biological imperatives of living organisms. The development of what was considered good science ultimately hinged on the contestation between these stakeholders' subjective gustatory and olfactory sensory experiences and their interactions with an unruly material. Ultimately, science lent value and authority to some stakeholders' sensory experiences over others as all searched for the perfect farm-raised catfish.

Prior to the rise of industrial fish production, the species' place in culture was clear. Scholars elucidate that both white and black southerners consumed the cat, but that African Americans became "particularly associated with the whiskered fish."[8] For instance, David Cohn famously identified the geographic Mississippi Delta as beginning in the lobby of Memphis's Peabody Hotel and ending in Vicksburg's Catfish Row, an area associated with African Americans. Moreover, George Gershwin's "Porgy and Bess" also pushed Catfish Row and "blacks' link to catfish into the national consciousness."[9] Food scholar Adrian Miller explains how the fish's reputation as a muddy river dweller and its ostensible flavor hitched the fish to racial stereotypes. "Life in mud also gives its meat a distinct muddy taste, creating a sharp dividing line between those who preferred the taste and those who detest it. So, while this catfish prejudice undoubtedly had an ugly racial tenor, it was also due in part to the fish's muddy taste, which turned off a lot of white consumers," Miller wrote.[10] Miller's description of catfish as "muddy" itself connects race, filth, and taste. But moreover, Miller's own acceptance of "muddy taste" as an empirical reality may be one we wish to historicize because clearly, for some Americans, the apparent preference of some (nonwhite) consumers for muddy flavors reinforced the notion that race was a truly meaningful difference, and that African American and white palates were, in fact, distinct. The nonfishy, neutral-flavored farmed catfish that farmers desired signaled their belief that the crop's flavor was the exact quality that could make it more appealing to consumers across race, class, and region, but specifically to higher-income, white consumers.

Ostensibly, neutral science, enacting the processors' and researchers' own subjective ideas of the most marketable catfish flavor, pulled the fish from the muddy depths of poverty and blackness and signaled a measured erasure of its racial and class ties.

Through processing technology, Miller observed, "Catfish farmers had actually succeeded in removing the muddy flavor. The aquaculture farmers developed a grain feed that gave catfish a uniform flavor, taste, and texture . . . the catfish's new, bland flavor also made it a good substitute for more expensive firm, delicate fish like sole, flounder, cod, sea bass, red snapper, and halibut."[11] While Miller's assertions echo other southern foodways scholars, the pursuit of processors' and researchers' desires, many of whom were white men, and the development of the technology and science to produce the constructed farm-raised flavor, have been left unexplored. An ideological reconfiguration accompanied the catfish's makeover from strong-tasting and wild to bland and domesticated. Through extensive quality-control measures and marketing, the industry transformed and slowly washed away the negative connotations that the wild fish's class and environment, and racial associations tethered to poverty, subsistence, and recreation.[12] The fish's industrial makeover had much to do with researchers', farmers', and processors' desires for a specific flavor that embodied a cleaner, blander, and whiter flavor.[13]

In seeing the farm-raised catfish as technology, we can understand how the pursuit of the bland farm-raised flavor traverses the histories of science and technology, animals, agriculture, environment, food, and the senses.[14] Most scholarship on the development of the farm-raised catfish industry focuses on its broad technological changes, and the management of flavor is an afterthought in those narratives.[15] Other works on the farmed fish's cultural history acknowledge the transformation from its wild to agricultural image, the fish's new taste, and the gentrification of its image.[16] Yet producers' systematic work to manage the fish's flavor, the implications of technology devised by researchers to regularize flavor, and the obstacles encountered from the animal and the environment are left largely unexplored. As these historical actors grappled with the notion of the ideal catfish taste, which validated and justified research conducted by land-grant scientists, their struggles against each other show how the actions of living organisms and interactions in their environments directly connected to and caused conflict among human actors. These actors contended time and time again with the catfish as a material that was once a living creature-now-turned piece of meat.

THE CHALLENGES OF MASS-CATFISH PRODUCTION

The pond-raised fish proved to be a lucrative endeavor. In the 1950s, experimental farmers in the South tried their hands at aquaculture. By the 1970s, the industry vertically integrated and the Yazoo, Mississippi, delta became the center of production.[17] John Egerton described the farm-raised catfish industry at this point as a "model of quality control and efficiency." He further claimed that it was a "rare anomaly in the food world: an artificially developed and mass-processed packed food that tastes better than its 'natural' predecessor."[18] Pond production made the catfish readily available all year long, regardless of consumers' time for recreation or proximity to a waterway. Prior to the advent of the industry, wild catfish consumption was based on localized tastes. Backed by intensive marketing and availability, by the 1980s, the catfish's image transformed. Catfish transformed from a food that large-scale producers saw as fit for the poor and those lacking keen sensibilities to one appreciated by many regardless of class, race, or location. Despite the rapid rise and success of the industry, farmers confronted innumerable challenges, especially in regard to flavor quality.

As catfish ponds began to proliferate across the South through the 1960s, the issue of flavor quality was for many purveyors initially inconsequential. Early fish farmers thought good taste was a given. As long as they fed their fish grain-based pelleted feeds like those used to raise other livestock, producers thought that the catfish would take on the feed's flavors. Yet feed alone could not create the perfect taste. Farmers did not take into account the catfish body and behaviors, and quickly discovered that the enclosed pond and pelleted food were simply not enough to ensure good flavor. In fact, the enclosed pond environment created a higher chance for off-flavors and intensified the problem as the growing catfish circulated with detritus, algae, and dead fish.

In the developmental years of the industry, grocery stores occasionally sold a fish without the knowledge of either the producer or consumer. In the 1960s, luckily for the producers, early pond-raised catfish consumers were, as Karni Perez observes, "accustomed to the slightly muddy flavor of the wild fish."[19] They were apparently indifferent toward any improved taste. And, in fact, they may have enjoyed those diverse and stronger flavors, not a consistently mild neutral flavor. Notwithstanding, farmers continued to claim that their crop had a different flavor profile from its wild counterpart.[20]

In order to achieve reliable market standing, the industry needed to create a sense of value in what had long been considered by many consumers to be a "trash" fish, and producers found that value in a closely controlled approach to flavor. As the editor for a catfish farming newspaper declared, "Quality and flavor—these are the keys to the industry's future growth. Farm-raised catfish must be sold on the basis that it is an agricultural product—produced with the same care, expertise and quality of other livestock. The quality and flavor, of course, distinguish the farm-raised product from river catfish and imported catfish." The editor continued, "Never–never for one moment should they be sacrificed."[21]

A consistently bland taste was vital to the growth of the farm-raised catfish industry, though not easily achieved. In the late 1960s, off-flavor became a pressing issue as farmers intensified their farming techniques and tried to expand their markets beyond traditional consumers. It was imperative that farmers put a product on the market that was free of flavors that the majority of US shoppers negatively associated with the wild fish's diet, behaviors, and environments. "Muddy" and "earthy" essences in farmed catfish could tarnish the "reputation of a successful fish farmer" and had the potential to give the "industry a black eye" if sold to the unsuspecting shopper.[22] As one industry booster claimed in 1971: "There is no doubt that producers and processors as a whole are already aware of the problem and its ramifications. Occasionally, however, a bad lot of catfish slips through, and the fine image that so many people are working so hard to improve, is tarnished."[23] When the industry promoted the crop as completely different from the wild animal, customers expected a "sweet [and] non-fishy" flesh.[24] The alleged difference between the wild and the industrially controlled fish was so great that to describe the fish as nonfishy seemed proper. But more, this desired flavor and supposedly positive characteristic of farmed catfish demonstrated that the industry wanted to sell their crop to a part of the public that did not, at this juncture, like seafood and preferred land-based animal flesh.

In the late 1960s through the early 1980s, the United States Department of Agriculture and the Marketing Research Corporation of America, among other agencies, determined the racial, class, and regional makeup of seafood consumers. In the early 1970s, consumers in the United States, particularly white and middle-class, were, compared to people of color and the lower classes, less likely to consume fish and seafood.[25] But within the decade, white middle-class dietary habits were changing. It became clear: more whites with higher incomes and education levels were eating more fish and seafood. One economist observed that the growing popularity of seafood could have been attributed to concerns over health.[26] While consumers with higher incomes and, perhaps, with educated preferences desired a more healthful diet, catfish farmers saw taste as a way to grow their consumer base. With consumers' attention turning anew to seafood, catfish farmers had a nonfishy fish to sell and they were ready to tap into the lucrative white middle- and upper-class market.

Wild catfish flavors embody the environments and waters in which fishermen catch their prey and those fish take on the flavors of the insects, other fish, and plant matter that the animals consume. "Catfish obtained from the wild sometimes possess a strong odor or taste reflecting the environment from which they were taken," researchers have claimed.[27] The farm pond also encloses water, plant life, and insects that could create the same wild gustatory qualities. Thus, the pond's ecology created a great deal of uncertainty for the farmer's crop quality. The pond environment is in fact tremendously chaotic. Weather, water, chemical contamination, bacteria, and algae could all contribute to repugnant-tasting flesh in an industrial farm pond. The presence of algae could cause undesirable flavors, particularly in the summer months. Algae quickly grow in warm waters and release odorous compounds, particularly geosmin and 2-methylisoborneal (MIB), into the water. Thus the weather has its impact: the warmth and photosynthesis produced by the sun can potentially generate undesirable flavors in the fish. But what is more, a breeze, a gust, whatever typically cuts the density of summer heat, can also cause off-flavor. Agricultural chemicals, especially from nearby spraying operations, may drift over ponds and cause undesirable gustatory attributes in the crop. All these contributing factors to off-flavor can mingle in a catfish pond environment and intensify the off-flavors through the very nature of the pond itself as an enclosed space.

Moreover, the catfish itself, as a living being, caused flavor problems. The catfish's decisions, its survival mechanism, and its body worked against farmers' and processors' pursuit of a clean, tasteless meat. Channel cats are piscivory, which means that they eat other fish. They are omnivorous too. For fishermen, this meant that the catfish was an easy catch, with effective types of bait easily found. But the animal's indiscriminate appetite worked both for and against farmers. For one, the fish's proclivity toward pelleted food made it easy to rear in ponds. But channel catfish also devoured rotten matter. Catfish stocks nibbling away on their own dead could take on an off-flavor. Scientists further confirmed, against their own presumptions, "that channel catfish will consume significant quantities of filamentous algae."[28] The farmed fish's unruly, multifarious decisions posed one major source of off-flavor, but another was its body. Its gills and gastrointestinal tract could dash farmers' and processors' dreams of a bland nonfishy flesh. As a catfish's digestive tract processes algae and metabolizes the plant matter into energy, off-flavor-producing compounds congregate into the fish's muscular tissue, its meat. Fat also stores undesirable flavor compounds. The fatter the cat, the longer it retained any off-flavor it had acquired. Because processors and researchers wanted a specific flavor, they fought against the animal, the environment, and the mechanics of the aquaculture itself.

Throughout the history of the farmed catfish industry, the enclosed chaotic pond environment baffled scientists. For instance, experiments at Auburn University, conducted between April and October 1983, demonstrated real bewilderment.[29] The scientists studied the connection between climate and season and severity of objectionable flavors. Their studies contradicted previous work on the connection between soil alkalinity and undesirable aromas in farmed cats. Earlier studies had determined that heavy alkaline soils were more likely to produce off-flavored catfish compared to acidic soils. Even more confusing for researchers, blue-green algae and actinomycetes, two organisms known to produce off-flavors, were abundant in ponds with catfish that tasted on-flavor. The researchers aptly observed, "There was considerable variation among ponds with respect to off-flavor scores."[30] With such varying results, they concluded, "The off-flavor problem is apparently complex, and the organisms and environmental factors responsible for the production of odorous compounds are largely unknown."[31] The scientists vigorously continued their quest for the causes of displeasing flavors.

The extensive studies on displeasing-tasting farmed catfish were part of a larger body of scholarship on off-flavored fish. American and European investigators began to examine undesirable tastes in fish in the early twentieth century.[32] In 1910, French researcher L.

Leger conducted the first study on muddy flavors in rainbow trout. He blamed them on *Oscillatoria tenuis*, a cyanobacterium produced by blue-green algae. Some twenty years later, A. C. Thaysen, an English researcher, published a study examining why "the richest salmon rivers of the kingdom had been found contaminated with an 'earthy' taint." What perplexed the researcher most was that the fish's intestines "were free of mud and were, in fact, practically empty."[33] Thaysen found that actinomycetes produced "earthy" pungent odors in salmon. Actinomycetes are filamentous bacteria that nurture in warm ponds and forage on uneaten feed and waste produced by fish. Actinomycetes are soluble in water, ether, and alcohol, and are "volatile in steam." In a concentrated form, they create "a brown amorphous material with a penetrating manurial odour."[34] In small doses, actinomycetes produced a soil-like smell and taste. It took researchers decades to pinpoint the specific compounds that produced the smell.

In the 1960s, a few scientists at Rutgers University's Waksman Institute of Microbiology researched the organisms in tainted waters. These microbiologists discovered the particular compounds that haunted catfish farmers in the years to come. In a 1965 study, Nancy N. Gerber and H. A. Lechevalier used a fairly new method of the era, gas chromatography that separated substances through vaporization, to isolate the substance they called *geosmin*, a colorless and highly odorous neutral oil from various actinomycetes. Gerber and Lechevalier named geosmin for the Greek root *ge*, or earth, and *osem*, or odor, because it produced a soil-like smell and flavor.[35] Other scientists found that a variety of actinomycetes, other than the subjects that Gerber and Lechevalier had examined, also produce the odorous oil.[36] Researchers discovered that blue-green algae like *S. muscorum* and *Oscillatoria tenuis* produced the viscous substance as well. A few years later, the Rutgers researchers stumbled upon another compound, MIB, that produced camphorous, musty odors.[37] Geosmin and MIB, it emerged, are the chief causes of perceived off-flavors in farmed catfish.

Land-grant researchers also chased the mysteries of off-flavor in farm-raised catfish, taking up their traditional responsibility to lead research of use for regional commerce. By 1971, Auburn University was leading investigations in the causes of off-flavor. Alabama's land grant was home to the prolific Dr. Richard "Tom" Lovell, whom catfish farmers and industry boosters honored as the "chief investigator of the 'whatdunit' of the underwater world."[38] The elimination of objectionable flavors and aromas in large-scale production posed an entirely new set of issues that even Lovell did not initially understand. Lovell's speculation on the causes of off-flavor stemmed from his experience with pond culture and his knowledge of the literature on off-flavor in carp, salmon, and trout. In a survey conducted from 1971 to 1972, the Auburn researcher learned that roughly half of all catfish farmers in Alabama produced some sort of off-flavored catfish.[39] Certain sectors of the farm-raised catfish industry, processors and researchers, in particular, were more concerned with off-flavored crops than were farmers. Processors were the most concerned with detecting undesirably flavored fish because they sold their fish to wholesalers, groceries, and restaurants. Some researchers focused their attentions on explaining the causes and devising cures for bad-tasting catfish. The processors and researchers became the accepted experts on flavor, not the farmers. In fact, many catfish growers chafed at flavor evaluators' appraisals of their fish that, at times, affronted the farmers' abilities to grow fine crops. The processors' senses became paramount over farmers' senses. Those farmers whose catfish just could not live up to the standard of good-tasting farmed cats quickly found that they had no other choice but to leave the business.

Off-flavors continued to occupy Lovell for years. Eighteen months after his initial studies began, the scientist's article "Fight Against Off Flavors Inches Ahead" asserted, "Research hasn't yet developed a guide to combat off flavors, much less determine the exact causes ... progress is being made." Lovell did not have much in the way of cures, but the article maintained that farmers misunderstood the undesirable flavors in the fish. With limited cures, Lovell cautioned, "There is one step that growers and processors can take which will minimize the hazard of off-flavor," and asserted that growers needed to take seriously the notion that "catfish are very sensitive to absorbing obnoxious flavors from the culture environment." Farmers needed to understand that catfish were porous vessels, and the pond was invariably contributing to the quality of the meat. Further, producers had to recognize that flavor was imperative to the health of the industry. Some growers did not believe that undesirable flavors could hurt the industry, and credulous others questioned if off-flavor even existed. Lovell warned, "An unpleasant flavor on the market will do serious and irreparable damage to the industry." Reminding farmers of the temporality of undesirable flavors, he wrote, "These flavors can, however, be purged from the fish so that they can be marketable."[40] For many, however, waiting for fish to become on-flavor was a painfully slow process.

In many cases, to follow advice like Lovell's meant farmers had to place bad-tasting crop in ponds with fresh water so that the fish could flush the displeasing flavors from their bodies. A rate of depuration varies, but typically, patient farmers would wait for two weeks. If the farmer had a limited source of water, he could wait for nature to takes it course. The farmer could wait for rain. The undesirable flavors would eventually evaporate as odorous compounds produced from algae disappeared as either the algae died off or the weather cooled. Lovell observed, "The off-flavor eventually will clear up . . . although in many cases several months have been required."[41] Yet not all farmers could afford to take a "wait and see" approach to achieve a mild-flavored crop.

In the 1980s, investigators looked to biological controls and cures for the off-flavor plague. Researchers engaged in experiments with polyculture—a process in which two or more species are grown together—to provide a cost-effective and algaecide-free method to control off-flavor.[42] In 1982, Les Torrans and Fran Lowell at the University of Arkansas at Pine Bluff reared blue tilapia and channel catfish together, and their findings were promising. The tilapia cleaned up the ponds and fed on two known contributors of off-flavor: plankton and detritus in the waters. Despite what at first looked like a boon, Torrans and Lowell observed, "There are a number of practical constraints to the successful application of this technology." First the tilapia sexually matured faster than channel catfish, and the researchers found that it would be difficult to capture just the filter feeders. But even if they could seine out the fully-grown tilapia, consumers' lack of knowledge and marketing posed "the major constraint to tilapia foodfish production." Because of this, Torrans and Lowell did not continue their studies the following year.[43] The aquaculture specialists studied biological and chemical controls, but realized that the most responsible approach to washing away undesirable catfish flavors remained time and perseverance. Throughout the industry's history, investigators continued their studies on the cures of off-flavor at the pond level.

THE PROBLEM OF PROBLEM DEFINITION

As some land-grant researchers studied cures, others tried to help the industry reach consensus on what was even considered a desirable or objectionable flavor, what was on- and off-flavored.[44] Processors and growers bickered about farm-raised catfish flavor. Published in 1974, a Southern Cooperative Series bulletin on catfish aquaculture prepared by Lovell and food technologist G. R. Ammerman revealed these tensions. The bulletin claimed that "catfish farmers are now generally aware of the off-flavor problem and are in position to appreciate the processor's evaluation of the flavor of fish that are to be processed." The bulletin revealed, however, that "disagreement between the two on this subject is not completely a thing of the past."[45] The researchers' observations indicated a lack of cohesion in the industry regarding the ideal farm-raised catfish flavor. Further, some farmers may have legitimately thought that their crop tasted good because it embodied the flavors of catfish they were already familiar with, the ones in the wild. Others, however, thought that processors held personal vendettas against some farmers. The researchers warned, "Do not process off-flavor fish. . . . It is important that the producer understands this and appreciates the fact that off-flavor is a serious and realistic problem and not a processor's excuse for not accepting fish."[46] Farmers themselves had to learn that the industry needed to manufacture a consistent flavor.

Taste evaluators at processing plants required training too. Lovell and Ammerman specified that inspectors had to be acquainted with strong-tasting catfish flavor, and that "it is difficult and precarious to evaluate fish for off-flavor unless the evaluator is familiar with this quality." For processors, novice flavor testers at their plants could be just as problematic as a disgruntled farmer. Even the evaluators could be unsure and unfamiliar with the varieties and intensities of undesirable smells and flavors. They too could be unsure of the ideal farm-raised catfish taste. The researchers suggested that rookie inspectors "should have a control fish for comparisons" and that "fish with no off-flavor and fish with distinct off-flavor should be kept on hand (in frozen storage)." Onsite samples fulfilled another purpose. "These control samples are also useful in demonstrating to a doubtful farmer that his fish have off-flavor," Lovell and Ammerman proposed.[47]

In service of this highly scientized detection and calibration of fish flavor, the carcasses underwent preharvesting rituals before processing that involved specific ways to dismember, cook, and smell the flesh. These preprocessing formalities still did not abate the highly subjective nature of flavor and smell; they confirmed subjectivity. A grower brought a fish from a pond ready for processing, and the evaluators had to decipher if the pond was ready by testing a sample. Inspectors dismembered the fish, wrapped the samples in aluminum foil, and steamed them in a double boiler. When the evaluators cooked the samples they would smell "the head space vapor when the container is initially opened" and then sample "the flesh very close to the bone from areas near the tail and at the anterior end of the carcass."[48] The process entrusted to testers a standardized means of evaluating fish for what was perceived as on-flavor or off-flavor. Formalizing and standardizing the process of flavor evaluation lent trained processors' senses authority over farmers' senses. The tests reified the evaluators' senses as the best for detecting off-flavor. With a standard test and a standardized tester, evaluators nosed out a specifically bland, nonfishy flavor. Lovell and Ammerman placed an onus on the testers and wrote, "The processor should feel an obligation to his customers and producers to conduct a thorough and precise evaluation of each pond of fish."[49] To further standardize the process of checking catfish flavor, processors looked to technologizing and purifying the testing experience. Joe Glover of STRAL directed his concerns toward a machine that could reproduce the same cooking conditions time after time. He discovered that the microwave was the exact tool for such a purpose. The microwave was felt to further standardize and technologize the flavor-testing process that was fraught with human error and subjectivity from its inception.

In the mid-1970s, although the microwave became a tool that catfish processors used to combat displeasing flavors, the human palate remained essential. The prescreening rituals remained largely the same. Evaluators taste tested a fish from a pond ready for processing. They dismembered the catfish and microwaved a piece without seasoning of any kind, even salt. Then they tasted the cooked fish. If the expert taste testers found the fish to be off-flavor, farmers had to wait a few weeks. As a consequence, flavor tasters at catfish processing plants became ever more indisputable as gatekeepers of the perfect-farmed flavor. The taste tester made few friends among the farmers, and many farmers accused the testers of bias and discrimination as their methods and devices became standard in the industry.

Some tester's abilities within this heavily constrained procedure became legendary. In the early 1980s, Delta Pride Catfish, a farmer-owned cooperative, hired Stanley Marshall. Marshall eventually became known as having "a million-dollar tongue" because he was so sensitive to ostensibly off-flavored catfish.[50] Yet his palate may have been considered too discerning for the practical aims of large-scale commercial fish production. Marshall, along with other taste testers, may have been the flavor gatekeepers, but farmers and researchers found that the testers' olfactory and gustatory sensitivity cost the industry, and farmers and processors continued to contest the farm-raised catfish flavor. In 1990, the Agricultural Cooperative Service (ACS) observed, "This subjective testing has presented a number of problems to the industry. Testing can be too severe or too lenient. Strict testing can be construed as a way for a processor to discriminate unfairly when choosing which farmer's fish to accept or not to accept any fish." More troubling, the ACS continued, "Lenient testing can be construed as a way for a processor to pay a lower pond price than the more strict processors."[51] In some cases, farmers felt that flavor evaluation had nothing to do with their crops or with documented consumer preferences; rather, it was an economic weapon wielded against them. Farmers' criticisms were not categorically paranoid accusations because when processors did not take their crops due to flavor, farmers lost money and time. The situation had real financial implications for the industry as a whole.

Processors' pursuit of the perfect catfish flavor hindered the efficiency of the industry. In 1992, Louisiana State University (LSU) food scientists L. S. Andrews and R. M. Grodner conducted consumer surveys to determine a standard for consumers' off-flavor tolerability. The food technologists observed that human quality controls, like Stanley Marshall, periodically rejected up to 90 percent of the fish they sniffed, rolled across their tongues, and then spit out. The LSU researchers' investigation centered on the sensitivity of professional taste testers and consumers, and they sought to gauge the professionals' sensitivities to displeasing flavors.[52] Indeed, the land-grant researchers argued that the professionals were too strict and that they could hurt business. "With this high rejection rate based on off-flavor, processors have not been operating at peak production and consequently have lost man hours and money," the LSU researchers claimed.[53] Moreover, the study found that there were in fact acceptable levels of off-flavor in consumer preference. "It is evident that the current standards of this processor's taste-testers were much more stringent than the consumer panel required and even preferred," the researchers concluded.[54] The study not only revealed that taste testers could decrease the efficiency of the industry, but that acceptability, to a certain extent, was a subjective contested trait between professional catfish tasters and the catfish consumers.

Researchers used both traditional scientific measurement devices and the human body to decipher acceptable flavors in farmed catfish. They used typical laboratory methods like gas chromatography to measure amounts of geosmin and MIB in samples, setting those against acceptable tolerances for good flavor. As Anthony Acciavatti has written on taste-centered industrial sciences of this period, researchers used machines to measure subjective qualities and turned them into objective measurements when determining quality. Also like Acciavatti's historical actors such as Robert Allen Boyer, researchers' bodies could determine what was right and wrong.[55] The nose that enclosed its mucus membranes and filamentous hairs, the mouth that encased the tongue and its papillae, became the contested sites of power over olfactory and gustatory qualities of the fish crop. Some researchers such as Lovell believed subjective senses: taste and smell could be developed into a "satisfactory objective test for off-flavor."[56] Although researchers could measure the amount of off-flavor in the crop, and set sensory standards and thresholds, testers' sensitivity and informed subjectivity still mattered on evaluation panels. Yet neither instruments, nor the testers' sensory perceptions, were adequate in identifying all the flavors that a catfish pond could produce, especially as the problem grew in complexity with each new investigation. Until the 1980s, most researchers focused on the typical "muddy," "musty," and "earthy" objectionable flavors. The concentration on those objectionable flavors determined research directions, and processors and researchers did not devise an official lexicon for what was on- and off-flavors until 1987.

THE PROBLEM OF PRECISION

The ubiquity of so-named muddy or earthy musty flavors long inhibited research on other undesirable flavors. In 1983, despite earlier instances of rare aromas and tastes in the crop, Lovell and other researchers finally formally recognized and categorized "new" catfish off-flavors. Over a sixty-day study, Lovell and his crew gathered and tasted fish from 220 commercial ponds in Alabama, Mississippi, and Arkansas. Twenty-four ponds produced unmarketable catfish. The researchers gathered a sensory panel composed of six experienced evaluators. Their catfish-savvy palates were shocked by the researchers' samples. Only twenty-five percent of the fish they tested were perceived to be muddy, earthy, or musty. The other seventy-five percent of the fish had rarer flavors or tasted nothing like anything they had previously encountered in farmed catfish. The characters they detected ran the gamut from staleness to notes of sewage, which were "the most subtle and harder to identify."[57] After the assessments, the panelists created descriptions for each, deliberated, and then came to a consensus. They not only devised terminology but quantified the intensities of the new flavors. On a scale of whole numbers, between two and ten, the panelists described ten as no off-flavor and two as extreme.[58]

Lovell and his research team unearthed and described a

smorgasbord of off-flavors in farmed catfish that they described as both etiologically anthropogenic and "natural." The panelists described one as a "fecal-type flavor" and another as like "a lagoon with large amounts of organic decomposition." Sewage was the most frequent. Evaluators described the second most-recurrent flavor as "stale" and "severely lacking freshness," which was a combination of many displeasing flavors. They also encountered the typical and familiar earthy and musty impression, which they described as "sharp, pungent, to algae-like to muddy." Other less-frequent but nonetheless problematic characteristics demonstrated the range and variety of the undesirable: rancid, metallic, moldy, and "cobweb."[59]

The "new" flavors were not necessarily new in the array of reactions recorded to particular specimens, but rather, they were officially recognized after years of what little attention previous researchers had paid them. As early as 1971, Lovell had noticed unusual off-flavors in farmed cats, and encountered fish that he observed "tasted like they came from a river just below where industrial affluent emptied." Yet he solely focused on the earthy-musty, or the "generally accepted terms in the literature. . . . It is the predominate type of off-flavor compound in catfish." As Lovell confronted a variety of catfish flavors at that juncture in his career, he concluded, "So there are still a lot of mysteries."[60] More than a decade later, Lovell conducted a full-blown study of these other flavors. "These off-flavors are not new," Lovell observed. But he justified the earlier absence of minor off-flavor studies, "because [these off-flavors] are more subtle and not as distinguishable as the earthy-musty, they have gone unrecognized or not been considered discriminatory."[61] The researchers' earlier disregard for ancillary flavors demonstrates both the evolving subjectivity of what was considered off-flavor and that the descriptions of off-flavors continued to became more complicated. The researchers' aims toward precision for flavor descriptors came under greater scrutiny as more research on off-flavored catfish continued and became more complex as well. More than anything, this study demonstrated that the taste testers desperately needed a standardized language to discuss and describe, and through such efforts to intervene in, farm-raised catfish flavors.

The industry lacked cohesion in one of the crop's most important elements: its flavor. A study conducted in 1986 marked an important turning point for the industry. That year in New Orleans, Louisiana, Peter Johnsen and other researchers at the Food Flavor Quality Research Division of the Southern Regional Research Center, which was part of the United States Department of Agriculture (USDA), developed a "lexicon of pond-raised catfish flavor descriptors."[62] For two days, influential and decisive figures in the industry—scientists, extension agents, industry representatives, and taste experts—"trained" their palates and minds, and fabricated a standard lexicon for flavor descriptors of on- and off-flavors.[63] The investigations and the subsequent sensory panels reveal how researchers sought neutrality through quantification and group consensus, and it was this small group of individuals who devised industry standards for the inherently idiosyncratic. Despite the absence of an industry-wide vocabulary before 1986, Johnsen observed, "The skill and training of individuals responsible for this task varies but, to date, they obviously have been successful." The researcher cautiously continued, "However, as individual businesses grow and the industry expands and matures, there is a need for some standardization of quality control practices to ensure both flavor quality and product consistency."[64] To create these basic quality control standards, they had to create the common vocabulary. To do so, the group learned "descriptive analysis." The process described by the American Society for Testing Materials manual on sensory tests is a "sensory method by which the attributes of a food or product are identified and quantified using human subjects who have been specifically trained for this purpose."[65] Through the practice of descriptive analysis, the distinguished group practiced creating cohesion by first devising a standard language for a variety of grape drinks and fish, but not catfish. [66] They standardized and calibrated their palates and minds.[67]

Then, and only then, were the panelists deemed ready for catfish. Through much deliberation, and breaks between each sample in which panelists cleansed their palates with crackers and spring water, the panelists generated three overarching descriptive areas, including aromatics, tastes, and feeling factors.[68] The subjective and sensitive human tongue of each panelist created the industry's standard for catfish flavor descriptors. It was finally with this study that the industry devised ways to describe what was on-flavor too. It was nutty, chickeny, and corny. But even too much of these desirable flavors could produce an off-flavor product. Martine van der Ploeg, an off-flavor catfish flavor researcher, observed in 1992, "Note that although these descriptors are considered positive flavor attributes, if chicken, corn, or buttery flavors dominate the mild catfish flavor, [the] fish may not be acceptable to a processor."[69] Regardless, Johnsen's 1986 study formulated a language to describe catfish characteristics based on panelists' sensations of taste, olfaction, and touch.[70] In configuring a standard lexicon, the industry stakeholders' bodies became tools and agents against off-flavors.

The standard lexicon of catfish flavors created new problems. Regardless of the training, discourse, and consensus that flavor evaluators underwent to create the standard vocabulary, Johnsen noticed that flavor evaluations still lacked accuracy and consistency. The evaluations needed objectivity, and Johnsen recognized the flaws in human quality controls. A few years after his lexicon study, the food technologist directed an investigation on the reliability of sensory evaluations for farm-raised catfish. Johnsen complained that previous studies made "no attempt to determine the precision and reliability of the evaluation[s]." Johnsen interviewed and selected participants based on a variety of stipulations related to taste, lifestyle, and commitment. Johnsen and his research team

Figure 1. "Flavor descriptors commonly used by fish taste panels in testing preharvest pond-raised catfish," in Martine van der Ploeg, *Testing Flavor Quality of Preharvest Channel Catfish Southern Regional Aquaculture Center Publication 431* (Stoneville, MI: Southern Regional Aquaculture Center, November, 1991), 3.

needed standardized testers so they could standardize catfish flavor-testing techniques. They chose sixteen nonsmoking participants who devoted a year to the study. Their palates had to be sensitive to catfish off-flavors. But of equal importance, the participants had to be able to effectively communicate, possess basic knowledge of flavors, and understand as well as recount chemosensory experiences. The panelists, ranging from ages nineteen to seventy-four, trained for seventy-five hours over a five-month period. They became familiar with descriptive analysis and a variety of fish descriptors. During the testers' meetings, they discussed, debated, and then created the very terms for a sensory ballot. To ensure that all panelists were on the same page, the researchers attached scores for each attribute and calculated a mean score for each. If an individual panelist's score deviated from the rest, they "were coached to improve performance."[71] Taste testers had to be standardized.

The standardized testers needed to test standardized testing objects too. Johnsen and other researchers held significant "concern over the performance capabilities of individual panelists and the panel as a whole, as well as, the material being evaluated."[72] For Johnsen and his crew, people were only half the problem, and the materiality of catfish bodies posed another. Using a technique that the research team called blended individual fish samples (BIFS), they pureed multiple samples of flesh in a food processor. The fish's body comprised a range of flavors, which depended on whether a sample came from anterior and posterior areas. More challenging, samples from the same pond could have inconsistent flavors too. The researchers blended various parts of multiple catfish to create samples. The BIFS was "more homogenous and thus better representative of the population," the team asserted.[73] The industry needed standardized testers and standardized materials.

Johnsen's studies revealed how groups reach consensus on subjective qualities such as flavor and smell. The subjectivity of human palate and nose, particularly those associated with science and production, constituted industry-wide thresholds that established off-flavors and on-flavors. Yet scientists continued in their quest for objectivity. They cast their eyes on machines.

By the early 2000s, researchers compared sensory instruments like the electronic nose to human taste testers. Their basic premise was that there were just too many problems with humans as gauges. "Current inspection of catfish quality relies upon sensory evaluation that can be subjective, prone to error and difficult to quantify," researchers argued.[74] Further, although taste testers could easily detect off-flavors, researchers argued that these inspections only produced "semi-quantitative data," and, moreover, that "humans readily succumb to sensory overload."[75] Due to the lack of hard numbers and the failure of human bodies, other parts of the food industry had already turned to electronic sensory devices. Geosmin and MIB, the substances that often caused displeasing flavors in farmed catfish, also caused problems in other products ranging from drinking water to paper tissue.[76]

Scientists found that machines and humans contested what was desirable and unpleasant. In a 2004 study, USDA researchers Casey Grimm, Steven Lloyd, and Paul Zimba of the Thad Cochran Warm Water Aquaculture Center in Stoneville, Mississippi, discovered that in relation to off-flavors in catfish, human faculties as agents and tools against off-flavor could be assessed against sensory machines. The researchers used electronic noses to "smell" their catfish samples and measured the amounts of geosmin and MIB. Professional taste testers chewed and rolled catfish samples on their sensitive fleshy palates, and then made their conclusions. Although machines and humans agreed on 76 percent of the samples, 24 percent of results remained in dispute. Either the machines found the samples to be off-flavor and the evaluators asserted the samples to be on-flavor or, more problematic, the instruments found samples to be on-flavor, and the testers disagreed. From the researchers' perspective, "the second disagreement is of greater concern as the instrumental method is considered to be more sensitive and to provide a greater level of objectivity."[77] As investigators adjusted the instrument's satisfactory thresholds for MIB and geosmin, human and machine still contested four pieces of catfish flesh. In regard to the contested snippets and the discrepancies between the two assessment methods, Grimm, Zimba, and Lloyd concluded, "The possible reasons for the disagreement on the four fish are unknown and could result from mislabeling, sample preparation error, and/or instrumental malfunction." In short, they concluded, "We have no definitive explanation for these four fish and consider them anomalies."[78] Sensory instruments, like human quality controls, could fail. As Anthony Acciavatti has written of Aaron Leo Brody's strain gage denture tenderometer, researchers used machines to mimic human sensations to test for quality but could not fully replicate the human sensory experience.[79]

Moreover, mechanical setups were expensive and economically unfeasible for some processors; perhaps not surprisingly, transcending both machine and human instruments, some land-grant scientists saw promise in animal technologies. Scientists considered animals with heightened and differing sensory experiences. In the early 2000s, Richard Shelby at Auburn University trained dogs to detect geosmin and MIB in water samples. He found "the dogs are as accurate" and "they're quicker" than testing waters for off-flavor-causing compounds or tasting the fish itself.[80] With this success, Shelby decided to test the dogs' abilities on processed catfish fillets. The team trained Rusty, a Labrador retriever mix, Maggie, a German Shepherd mix, Ralph, a setter mix, and Ginger, a chow mix, to sniff out off-flavors associated with geosmin and MIB in cat samples. On average, the dogs were found to be eighty-one percent accurate. Ginger was even more precise and scored a whopping 90 percent accuracy rate. Scientists found pitfalls with the canine inspectors, however. While human evaluators could easily detect what researchers' had predetermined to be "unique" and nasty, dogs might find the same flavors "agreeable, or even pleasant," and they would thus "not be identified as off-flavour."[81] The researchers concluded, "We do not propose that dogs replace humans as 'taste-testers' at catfish processing facilities."[82] Indeed, dogs had acute olfactory experiences that had the potential to detect off-flavor. But the dogs' subjectivity and their preferences for what they considered pleasurable and repugnant fell in line with similar obstacles that farmers, processors, and researchers experienced in relation to each other. Some just could not agree on what was good and bad-flavored farm-raised catfish.

CONCLUSION

By the 1980s, newspapers and cookbooks lauded the catfish's makeover. They compared the farm-raised to the wild and posited that the crop was exceptional. The wild fish "have a strong flavor, a muddy taste. They're poor folks' food," Merle Ellis claimed in 1987. "Today's farm-raised catfish are among the finest, freshest, most flavorful and versatile fish you'll find in any market or on any restaurant menu," Ellis concluded. The reporter underscored agricultural control, pristine environments, and the habits of the fish crop as to why it tasted better, and thus why the crop was better than its wild cousins. The author argued that the catfish's "bad rap ... resulted because 'natural' catfish, those that populate every river, stream and pond all down the center of the continent and across the South, are 'bottom feeders.'"[83] Food writers praised the manufactured fish's light flavor, and one claimed, "The flavor of catfish, which is as bland and inoffensive as that of tofu, makes them suitable for highly seasoned sauces."[84] Others stressed the farm-raised catfish's nonfishy flavor and compared it to another well-known grain-fed industrial food: chicken.[85]

Although the industry engaged in an extensive marketing campaign, it also ensured nonfishy bland meat entered the markets. In doing so, the character of catfish consumers changed. The bland, neutral flavor made the catfish anyone's culinary canvas; that standardization of universality made the catfish a whiter food. In the 1970s, catfish consumers were more likely to be poorer, less educated, and predominately African American. By the 1990s, scholars discovered that unlike decades prior, more consumers in the United States who were white and possessed more education and higher income levels were more likely to consume the fish.[86] These studies affirmed that farmers, processors, and researchers, albeit in conflict over what was considered good-tasting catfish, determined the farmed fish's new consumer base through their own understandings of what on- and off-flavor meant for the image of the crop.[87] In 1991, one economist found the most important perceptions of the catfish related to its flavor, absence of a fishy taste, nutrition, and socioeconomic variables. The economists argued that an effective way to alter attitudes toward the fish was to stress the improved fish's flavor and carefully controlled pond culture. Negative attitudes, the economists found, stemmed from attitudes toward perceived "wild" and "muddy" flavors.[88]

The catfish's redemption into a new material demonstrates how the constructedness and subjectivity of flavor informed constructions of race and class. Ensuring a mild fish hit consumer plates was meant to cater to what the industry thought white, middle-class, and upper-class palates preferred. The development of science and technology aimed at flavor quality shows that science was far from objective. Rather, processors' and researchers' perceptions of what was good, acceptable flavor was loaded with subjective notions of racial and class difference. Moreover, what farmers, processors, scientists, and shoppers considered displeasing or off-flavor was contingent on the individual, their goals, and their sensitivity. The industry changed the catfish into a blander and whiter food, but the arduous process was fraught with struggles between living organisms and the industry's key players and their senses.

The industry standardized the catfish body, catfish flavor, and the catfish evaluators who determined what was considered good-tasting fish. The search for the subjectively bland nonfishy farm-raised catfish was as burdened with contingency and chaos as the pond environment itself. The catfish caused uncertainty for the industry, precipitated research, and its flavors triggered disputes among farmers', processors', and researchers' palates. Indeed, for commercial purposes, the unruly living organism was a challenging material object to control and standardize. The farm-raised catfish fought back.

NOTES

1. Parts of this essay appear in Karen Senaga, "Catfish Image, Catfish Taste: The Land-Grant System and the Development of Catfish Aquaculture in Mississippi," *Journal of Mississippi History* 75, no. 1 (Spring 2013): 33–52. I use multiple terms for the farm-raised catfish, including catfish, cat, fish, pond-raised catfish, farmed catfish, farmed fish, and crop.
2. Karni Perez, *Fishing for Gold: The Story of Alabama's Catfish Industry* (Tuscaloosa, AL: Fire Ant Press, 2006), 13.
3. Perez, *Fishing for Gold*, 73.
4. Perez, 73. The quotes from Stephens are likely paraphrased, as told to Karni Perez.
5. Regardless of the subjective nature of taste and smell, researchers described off-flavors as "objectionable flavours [sic] and odours [sic] that affect natural and municipal water supplies, as well as commercial and native fish population." The typical off-flavor is labeled as muddy, earthy, and musty. See J. F. Martin, C. P. McCoy, C. S. Tucker, and L. W. Bennett, "2-Methylisoborneol Implicated as a Cause of Off-Flavour in Channel Catfish, *Ictalurus punctatus* (Rafinesque) from Commercial Culture Ponds in Mississippi," *Aquaculture and Fisheries Management* 19 (1988): 151; Richard Stickney, an aquaculturist, has described off-flavored fish: "Actually, they just taste like mud: they're nasty." See Robert Stickney, *Aquaculture in the United States: A Historical Survey* (New York: John Wiley & Sons, 1996), 237.
6. William Boyd, "Making Meat: Science, Technology, and American Poultry Production," *Technology and Culture* 42, no. 4 (October 2001): 631–64. Boyd's article references Rachel Carson to conjure imagery of how the animal body "fights back" against antibiotics and intensive agriculture.
7. Senaga, "Catfish Image, Catfish Taste," 47.
8. Anthony Stanonis, "Just like Mammy Used to Make: Foodways in the Jim Crow South," in *Dixie Emporium: Tourism, Foodways, and Consumer Culture in the American South*, ed. Anthony Stanonis (Athens: University of Georgia Press, 2008), 220; Adrian Miller, *Soul Food: The Surprising Story of an American Cuisine: One Plate at a Time* (Chapel Hill: University of North Carolina Press, 2013), 70–90.
9. Stanonis, "Just like Mammy," 220.
10. Miller, *Soul Food*, 75–76.
11. Miller, 79.
12. For more on class and consumer culture, see Lawrence Levine, *Highbrow/Lowbrow: The Emergence of Cultural Hierarchy in America* (Cambridge:

Harvard University Press, 1990); Lizabeth Cohen, *A Consumers' Republic: The Politics of Mass Consumption in Postwar America* (New York: Vintage, 2008).

13. Scholars have discussed the issue of blandness and its connections to the construction of whiteness; see Camille Begin, "Partaking of Choice Poultry Cooked a la Southern Style: Taste and Race in the New Deal Sensory Economy," *Radical History Review* 110 (Spring 2011): 128, 131.

14. On taste and the creation of industrial foods, see Gabriella Petrick, "The Arbiters of Taste: Producers, Consumers, and the Industrialization of Taste in America, 1900–1960" (PhD diss., University of Delaware, 2006); Warren Belasco and Philip Scranton, eds., *Food Nations: Selling Taste in Consumer Societies* (New York: Routledge, 2002). For more on industrial agriculture, particularly pertaining to animals, see Steven Striffler, *Chicken: The Dangerous Transformation of America's Favorite Food* (New Haven, CT: Yale University Press, 2005). On the industrialization of animals and plants, see Susan R. Schrepfer and Philip Scranton, eds., *Industrializing Organisms: Introducing Evolutionary History* (New York: Routledge, 2004). For works that have addressed the development of animals in the laboratory setting, see Anders Halverson, *An Entirely Synthetic Fish: How Rainbow Trout Beguiled America and Overran the World* (New Haven, CT: Yale University Press, 2010); Robert Kohler, *Lords of the Fly: Drosophila Genetics and the Experimental Life* (Chicago, IL: University of Chicago Press, 1994); Karen Rader, *Making Mice: Standardizing Animals for American Biomedical Research, 1900–1955* (Princeton, NJ: Princeton University Press, 2004); and William Boyd, "Making Meat: Science, Technology, and American Poultry Production," *Technology and Culture* 42, no. 4 (October 2001): 631–64. On the complexities of the interaction between nonhuman and human actors, see Bruno Latour, *Politics of Nature: How to Bring the Sciences into Democracy* (Cambridge, MA: Harvard University Press, 2004); and Isabelle Stengers, *Cosmopolitics*, vol. 1, trans. Robert Bononno (Minneapolis: University of Minnesota Press, 2010).

15. For overviews of the history of catfish aquaculture history, see Senaga, "Catfish Image, Catfish Taste"; John Hargraeves, "Channel Catfish Farming in Ponds: Lessons from a Maturing Industry," *Reviews in Fisheries Science* 10 (2002): 499–528; Richard Schweid, *Catfish in the Delta: Confederate Fish Farming in the Mississippi Delta* (Berkeley, CA: Ten Speed Press, 1992); and Perez, *Fishing for Gold*.

16. Richard Schweid, too, discusses the importance of flavor and the catfish industry; Schweid, *Catfish in the Delta*, 39–55.

17. On the vertical integration of the farm-raised catfish industry, see John A.

Hargreaves, "Channel Catfish Farming in Ponds: Lessons from a Maturing Industry," *Reviews in Fisheries Science* 10, nos. 3/4 (2002): 499–528.
18. John Egerton, *Southern Food: At Home, on the Road, in History* (Chapel Hill: University of North Carolina Press, 1993), 134.
19. Perez, *Fishing for Gold*, 210.
20. Perez, 210.
21. "An Editorial: The Pitfalls of Off-Flavor," *The Catfish Farmer News Leader* 2, no. 3 (November 1971): 2.
22. Roy Grizzell, "Off-Flavors in Catfish," *The Catfish Farmer*, Summer 1969, 10.
23. "An Editorial," 2.
24. "An Editorial," 10.
25. Tehwei Hu, *Analysis of Seafood Consumption in the United States: 1970, 1974, 1978, 1981* (Silver Spring, MD: National Marine Fisheries Service, 1985), 9.
26. Hu, *Analysis of Seafood*, 40.
27. W. Guthrie Perry and James Avault, "In Brackish Water: Catfish Culture Studies in Louisiana," *The Catfish Farmer*, March–April 1970, 22.
28. Robert T. Lovell and Lewis Sackey, "Absorption by Channel Catfish of Earthy-Musty Flavor Compounds Synthesized by Cultures of Blue-Green Algae," *Transactions of the American Fisheries Society* 102, no. 4 (1973): 777.
29. Martin S. Armstrong, Claude Boyd, and Richard T. Lovell, "Environmental Factors Affecting Flavor of Channel Catfish from Production Ponds," *The Progressive Fish-Culturist* 48 (1986): 113.
30. Armstrong, Boyd, and Lovell, 116.
31. Armstrong, Boyd, and Lovell, 116–118.
32. Per-Edvin Persson, "19th Century and Early 20th Century Studies on Aquatic Off-Flavours—A Historical Review," *Water Science Technology* 31 (1995): 10.
33. A. C. Thaysen, "The Origin of an Earthy or Muddy Taint in Fish," *Annuals of Applied Biology* 23, no.1 (1936): 100.
34. Thaysen, 103.
35. N. N. Gerber and H. A. Lechevalier, "Geosmin, an Earthy-Smelling Substance Isolated from Actinomycetes," *Applied Microbiology* 13, no. 6 (November 1965): 935.
36. Richard T. Lovell, "Flavor Problems in Fish Culture" (Rome: Food and Agriculture Organization of the United States, 1976), 459.
37. Nancy N. Gerber, "A Volatile Metabolite of Actinomycetes, 2-Methylisoborneol," *The Journal of Antibiotics* 22, no. 10 (October 1969): 508–9.
38. "Off Flavor," *The Catfish Farmer*, November 1971, 8.

39. Robert T. Lovell, "Fight against Off-Flavors Inches Ahead," *Fish Farming Industries* (February 1972): 22.
40. Lovell, "Fight against," 24.
41. Lovell, 22.
42. Les Torrans and Fran Lowell, "Effects of Blue Tilapia/Channel Catfish Polyculture on Production, Food Conversion, Water Quality and Channel Catfish Off-Flavor," *Proceedings Arkansas Academy of Science* 41 (1987): 82.
43. Torrans and Lowell, "Effects of Blue," 84.
44. On-flavor refers to farm-raised catfish that do not have any indication of off-flavors.
45. Robert Lovell and G. R. Ammerman, "Processing Farm-Raised Catfish: A Report from the Processing and Marketing Subcommittee of Project S-83," *Southern Cooperative Series Bulletin* 193 (October 1974): 37.
46. Lovell and Ammerman, "Processing Farm-Raised," 40.
47. Lovell and Ammerman, 42.
48. Lovell and Ammerman, 42.
49. Lovell and Ammerman, 42.
50. To read more about Stanley Marshall and flavor testing, see Schweid, *Catfish in the Delta*, 45–48.
51. David Wineholt, "Cooperative Builds Delta Catfish Industry, Brings Price Stability, Assured Market," *Farmer Cooperatives* 57, no. 4 (July 1990): 15.
52. L. S. Andrews, R. M. Grodner, and the Louisiana Agricultural Center, "Consumer Survey of Pond Raised Catfish to Establish a Standard Level of Flavor Acceptability," (unpublished research, LAES, Louisiana State University, 1992), 1.
53. Andrews and Grodner, "Consumer Survey," 1.
54. Andrews and Grodner, 3.
55. Anthony Acciavatti, "Do Industrialists Dream of Soy Sheep?" (manuscript in preparation, n.d.).
56. R. T. Lovell et al., "Objective Analysis of Fish for Off-Flavor," *Highlights of Agricultural Research* 33, no.1 (1986): 20.
57. Richard T. Lovell, "New Off-Flavors in Pond-Cultured Channel Catfish," *Aquaculture* 30 (1983): 329.
58. Lovell, "New Off-Flavors," 330.
59. Lovell, 331.
60. Lovell.
61. Lovell, 332.
62. Peter B. Johnsen, Gail Vance Civille, and John Raymond Vercelloti, "A Lexicon

of Pond-Raised Catfish Flavor Descriptors," *Journal of Sensory Studies* 2 (1987): 85–91.
63. Senaga, "Catfish Image, Catfish Taste," 47.
64. Johnsen, Civille, and Vercelloti, "Lexicon," 86.
65. Robert C. Hootman, *Manual on Descriptive Analysis Testing for Sensory Evaluation* (Baltimore, MD: American Society for Testing and Materials, 1992), 1.
66. To read more about descriptive analysis, see "Descriptive Sensory Analysis: Past, Present and Future," *Food Research International* 34 (2001): 461–71.
67. On the origins of standardization testing and materials, see Amy Slaton, *Reinforced Concrete and the Modernization of American Building* (Baltimore, MD: Johns Hopkins University Press, 2001).
68. Johnsen, Civille, and Vercellotti, "Lexicon," 88. The standard aromatics were nutty, boiled chicken, grainy, MIB, or a flavor associated with blue-green algae or geosmin, or a taste related to decaying wet wood, putrid, rotten plants, cardboard, and paint. The tastes were sweet and salty, while the feeling factors were astringent and metallic.
69. Martine van der Ploeg, "Testing Flavor Quality of Preharvest Channel Catfish," *SRAC* 431 (Stoneville, MS: Southern Regional Aquaculture Center: November 1991), 4.
70. Johnsen, Civille, and Vercellotti, "Lexicon," 86.
71. Peter Johnsen and Carol Kelly, "A Technique for the Quantitative Sensory Evaluation of Farm-Raised Catfish," *Journal of Sensory Studies* 4 (1990): 191
72. Johnsen and Kelly, "Technique," 190.
73. Johnsen and Kelly, 191.
74. Figen Korel, Diego A. Luzuriaga, and Murat O. Baiban, "Quality Evaluation of Raw and Cooked Catfish (*Ictalurus punctatus*) Using Electronic Nose and Machine Vision," *Journal of Aquatic Food Production Technology* 10, no. 1 (2001): 3.
75. Casey C. Grimm, Steven W. Lloyd, and Paul Zimba, "Instrumental versus Sensory Detection of Off-Flavors in Farm-Raised Channel Catfish," *Aquaculture* 236 (2004): 310.
76. Grimm, Lloyd, and Zimba, "Instrumental," 310.
77. Grimm, Lloyd, and Zimba, 316.
78. Grimm, Lloyd, and Zimba, 316.
79. Acciavatti, "Do Industrialists Dream of Soy Sheep?"
80. David Elstein, "Something's Fishy: Training Dogs to Smell Off-Flavor in Catfish," *Agricultural Research*, April 2004, 11.
81. Richard A. Shelby et al., "Short Communication: Detection of Off-Flavour

in Channel Catfish (*Ictalurus punctatus* Rafinesque) Fillets by Trained Dogs," *Aquaculture Research* 37 (2006): 301.
82. Shelby et al., "Short Communication," 300.
83. Merle Ellis, "Farm Raised Catfish Dispel Bad Reputation," *St. Petersburg Times*, March 26, 1987, 5D.
84. Nancy Harmon Jenkins, "Lessons from the School of the Unfamiliar Fish," *New York Times*, January 8, 1986, B-8.
85. Lad Kuzela, "Fish Farmers Get a Whiff of Profitability," *Industry Week*, October 4, 1982, 48; Senaga, "Catfish Image," 51.
86. Lynn E. Dellenbarger, Alvin R. Schupp, and Paula August, "Household Consumption of Catfish in Louisiana by Product Type," *Journal of Food Products Marketing* 3 (1996): 38, 43; Senaga, "Catfish Image," 51.
87. Senaga, 51.
88. Senaga, 26.

CHAPTER THREE

ROOM AT THE BOTTOM

*The Techno-bureaucratic Space of
Gold Nanoparticle Reference Material*

Sharon Tsai-hsuan Ku

INTRODUCTION

Standard Objects: The Ubiquitous Invisibles

Anyone who has experienced working in laboratories should be familiar with the following scenarios: The most quiet space is normally saved for expensive instruments; the highest quality refrigerator is used to store mundane experimental materials; all kinds of warning signs are posted on the walls reminding human beings to behave well, not to destroy or disturb these delicate occupants; and researchers often stay up all night to take care of lively cells on culture dishes or hidden atoms that refuse to appear under the microscope. This material culture of labs not only contributes to the triumph of experimental science by bringing in "universality" and "objectivity," but also constructs a unique relationship between humans and objects. Subjective human beings and their judgments are constantly removed from the system, allowing material objects to be produced and to replace the role of humans.

The production and application of standardized scientific objects provides a vivid illustration of this ambivalent boundary between the material and human world. A standardized scientific object has to possess three qualities. First, it has to be made by precise measurement and well-controlled manufacture. Take the production of the atomic clock, for example. To make a standardized atomic clock, a pure cesium source, well-calibrated instruments and accurate measurements to obtain the atomic oscillation frequency are all indispensable. These scientific requirements also have to be regulated by metrology, the science of measurement, indicating that this standard timekeeper is highly scientized. Second, a standardized scientific object is not just a product of exact science; it has to reflect social consensus and be used in human society. The atomic clock becomes a standard timekeeper because of the social agreement among particular communities to have a synchronized timeframe and, by dint of the collective action to broadcast its time signals by radio across the world. Third, for the standard to be widely distributed, these delicately constructed scientific and social infrastructures have to become invisible. The atomic clock serves as such a successful standard because it makes us "forget" all the complicated measurements and social commitments required; it enables our dependence on it to be as subtle and necessary as our natural breath.

These three characteristics differentiate standardized scientific objects from scientific objects, as the former must possess not only rigorous scientific properties but also mundaneness and routines to connect with ordinary social life. From this perspective, standardized scientific objects do not act as saliences that mark scientific progress. Quite the opposite. What is customarily labeled as the scientifically or industrially "destructive or novel" may in fact indicate continuities, adaptation, and consistency. In other words, their success comes from their ubiquitous but invisible existences. How can these seemingly contradictory characteristics—being thoroughly novel, scientific but extremely mundane, very technical yet completely social—coexist in standardized scientific objects?

Precision, Novelty, and Standardization: The Puzzle in Nanotechnology

The measurement of nanoparticles offers the best example to address this paradox. The entire field of nanotechnology is based on the assumption of "precision measurement" to capture the "novel physical and chemical properties" of particles at the range of nanoscale (1–100 nm) for new material application (a definition provided by the website of the US government's National Nanotechnology Initiative [NNI]). The history of nanotechnology is often traced back to Richard Feynman's 1959 lecture at the California Institute of Technology (Caltech), where he predicted that miniaturization would be the trend of future technology, offering "plenty of room at the bottom" to rapidly enhance information storage and process.[1] The canonical history attributes the scope-setting of nanotechnology to Norio Taniguchi, the first person who explicitly used the term "nanotechnology" and defined it as "the processing of, separation, consolidation, and deformation of materials by one atom or by one molecule."[2]

The narrative on atomic precision and exact measurements has been heavily mobilized at the early planning stage of nanotechnology research and application by the US government.[3] Mainstream nanotechnology discourse has created an imagery in which the heroic device, the scanning tunneling microscope (STM), and its inventors, two Nobel Laureates from IBM-Zurich Laboratory, Heinrich Rohrer and Gerd Binnig, serve as the iconic figures of nanotechnology. On January 21, 2000, then-president Bill Clinton delivered a speech at Caltech, the same place where Feynman had made his statement decades earlier, to announce the National

Nanotechnology Initiative. Since then, a field that advocates the idea that "size matters" has become by a number of measures the major field in US science and technology. The overall investment from 2000 to 2012 reached about $20 billion, with the assumption that atomic precision will lead to nanoscale innovation: the ability to see at the atomic scale through instruments such as the STM or atomic force microscope (AFM) has been considered by the NNI to be the starting point and essential foundation of nanotechnology, for the precise characterization, control, and manipulation of novel material properties.

The development of nanomedicine, a program launched at the National Cancer Institute (NCI), illustrates the taken-for-granted assumption of nanoscale precision as a universal standard for novel material breakthrough. Scientists have claimed that nanoparticles' tiny size and large surface area make it an ideal drug carrier that can efficiently transport drugs to targeted tumor sites (see figure 1a). However, precision, standardization, and novelty of size measurement ironically have been the most troublesome concepts ever since the announcement of the NNI, and have fired up endless conversations and negotiations within scientific communities and among various social institutions. The main funding agency, the NCI, had noticed that many laboratories that claim to use the same material as drug carriers reported inconsistent particle-size values and size effects. Different laboratories customarily perform experiments based on their own instrumental settings and measurement protocols; subsequently, size values obtained by one lab can rarely be reproduced by another. Moreover, currently the entire nanotechnology community struggles with basic characterization issues. Nanoparticles viewed under electron microscopes are often observed as aggregations coupled with unknown background noises from the sample preparation or vibrations, instead of atomic or elaborate functional structures. Instruments with atomic precision apparently cannot guarantee the accuracy of nanoscale measurements that are ten times larger than an atom.

This paradox challenges the linear assumption that angstrom-scale precision can automatically lead to nanoscale precision. How does one obtain a precise nanoscale characterization if it cannot be determined by instruments with much more sophisticated atomic precision? Furthermore, if the very concept of "atomic precision" is not a universal standard, why is the belief in "precision and accuracy" still deeply rooted in nanotechnology reasearch and development policy discourse? What ultimately justifies the scientific credibility of a nanoscale measurement?

To understand the relationship among standards, precision, and their social origins, this paper offers a sociological analysis of the production of the first national nanosize standard, gold nanoparticle reference material (gold RM) and its application to nanomedicine characterization. Produced by the National Institute of Standards and Technology (NIST), the gold RM was delineated as an objective "nano-ruler" to resolve the inconsistency of nanodrugs' size measurements. By putting it side by side with testing samples (see figure 1b), the gold RM was expected to calibrate instruments, protocols, and data processes used by different nanodrug developers, enabling them to obtain reproducible and comparable data.

The production of the gold RM brings up a critical aspect about the role of bureaucracy in precision measurement and scientific innovation. Being notoriously known for its rigidity and dependence on routine, bureaucracy seems to be a counter factor against nanotechnology as revolutionary science and destructive innovation. However, by contrast, this chapter argues for the complex interaction of the bureaucratic culture and material culture in constructing and securing the definition and value of nanoscale standards, emphasizing the following three aspects:

1. Nanoscale precision and its application in nanoscale innovation are legitimated and sustained, not by its scientific novelty but by the mobilization of bureaucracy. Without

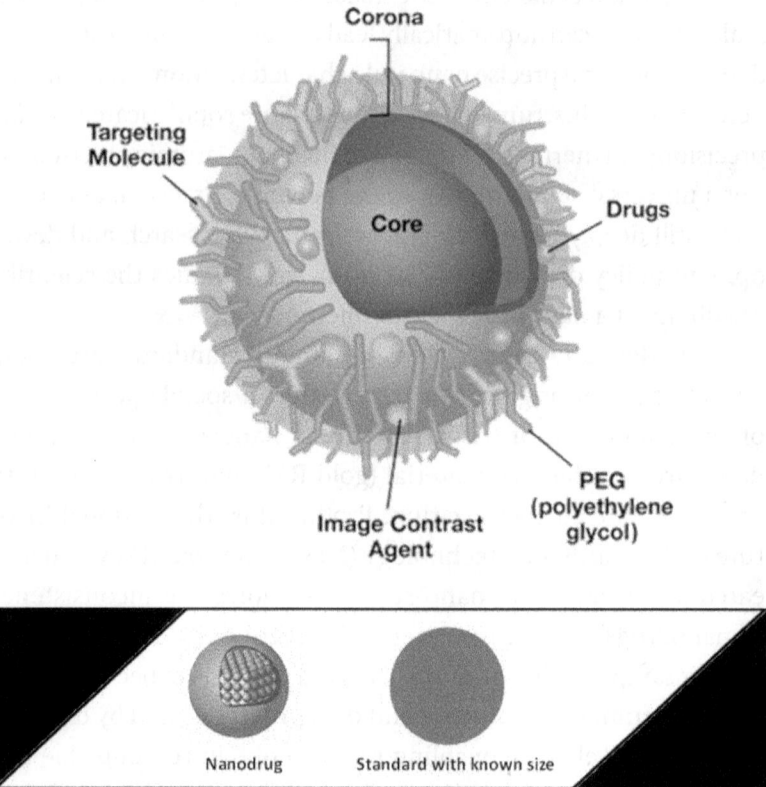

Figures 1a and 1b. Schematic diagram of a nanodrug; measuring the nanodrug using a size standard as a reference. *Source*: Nanotechnology Characterization Laboratory.

the activation and operation of the bureaucratic order to manage scientific information, precision and standard measurement cannot be achieved.

2. Atomic precision cannot guarantee a trustworthy nanoscale standard. The concepts of size, precision, and standards are not self-evident. Instead, they need to be contextualized in the institutional histories of NIST and NCI, as well as the disciplinary hierarchies between physics and biology in basic science research.

3. Rather than harmonizing scientific controversies, the nano-size standard has become the source of controversies that demand social means to close the debates.

Through tracing the coming-into-being of the gold RM, I develop the techno-bureaucratic object as a theoretical concept to analyze systematically how bureaucratic mundaneness, indifference, and rigidity contribute to material robustness, data harmonization, and metrological precision, as well as the dynamic mutual shaping between epistemic/social contingency and nanotechnology standardization.

Literature Review

Standardization has been a critical lens for STS scholars and historians of science, examining the strong entanglement of scientific, national, cultural, and economic values.[4] These deeply buried social factors in the production of standard objects challenge the meaning of precision and its relationship to standardization, where local variations, contingencies, step-by-step traceability, and subjective practices are necessary and essential conditions in the making of universality and objectivity.[5] These bottom-up practices further indicate standardization as social institution, where organizations such as bureaucracy play a critical role in generating trust in numbers. The "machine-like bureaucracy," as argued by Theodore Porter, corresponds with Bruno Latour's proposal of viewing standardization as "center of calculation," constituted by a collection of "immutable mobiles" such as writings, inscriptions, and documents, which can move in time and space while being interpreted the same way in various contexts.[6] It is through such operations that these ubiquitous immutable objects seamlessly and incrementally translate the uncertain real world into a series of controllable laboratories, where compatibility, standardization, and universality are gained.[7]

While the existing literature offers a solid foundation for analyzing the standardization of new and emerging technologies, a more complex and dynamic view about the "social interests" involved in the process of standardization is needed, given increasingly heterogeneous private-public partnerships and organizations involved in contemporary standard-making. For example, the triple-helix alliance among government, industry, and academia has complicated the roles of the state and private sectors in standardization, raising critical challenges about whether or not standardization should serve for regulation, innovation, or commercialization.

Similar questions appear in the interpretation of bureaucracy. Both Porter and Timmermans (who carry distinct views on the role of bureaucracy in standardization) assume that bureaucracy functions as a rational machine producing "impersonal knowledge." However, anthropologists have demonstrated the importance of informal organizations in formal bureaucratic processes.[8] These delicate and subtle practices within bureaucratic organizations are crucial in contemporary science and technology standardization, as many standardization activities are technical, commercial, and regulatory in character, and require the participation of multiple agencies that possess distinct bureaucratic structures and cultures. (The nanosize standard discussed in this paper is an example.) In other words, bureaucracy is not a homogeneous entity in the context of technoscience standardization. To be able to differentiate different bureaucratic practices, and how they contribute to the formation of national certified standards, requires further clarification.

Last but not least, there is the problem of materiality. Most of the current literature treats standard objects as a black-boxed technology, thus the analytical strategy is to open the black box to recover the social process of standardization in the making. However, for standard objects created for emerging technologies, there is no such material certainty, even for those objects listed in textbooks. In other words, there is a gray line between ordinary standards and innovative science that analysts have to be aware of when analyzing standardization in advanced technology. Cyrus Mody and Michael Lynch propose the idea of the "test object" to illustrate the material uncertainty of standard objects in emerging technology.[9] Using the Si (111) 7 x 7 surface, a widely used scanning tunneling microscope (STM) calibration object as the case study, they argue that a test object has variable shadings of practical, mathematical, and epistemic significance allowing it to travel back and forth between laboratory routines and research territories, being either a calibration device or research subject to satisfy different users' interests. Mody and Lynch's argument challenging standards for newly emerging science and technology as mature black-boxed objects is well-taken. Yet, the freedom they grant to these objects, allowing them to travel back and forth between research frontier and mundane laboratory routine, seems to miss the very criterion of standards being a "center of calculation." In the context of national standardization, their argument requires more systematic investigation.

Techno-bureaucratic Objects

I develop techno-bureaucratic objects as the conceptual framework to study the relationships between science, bureaucracy, and the gold RM's materiality. My intention is to overcome the uneven treatment in current standardization literature that often devotes all the analytical attention to scientific practices, using bureaucracy as a context or explanatory factor, rather than as an "actor" whose participation in science also deserves to be explained. The techno-bureaucratic object gives equal attention to actors' symmetrical treatment of scientific and bureaucratic practices when materializing standard objects. These two forms of practice go hand in hand, and often they mutually stabilize each other in the process of standardization.

I use both the boundary object, a concept developed by Susan Leigh Star and James Griesemer, and document analysis to conceptualize the techno-bureaucratic object. According to Star and Griesemer, boundary objects have different meanings in different social worlds, but their structure is common enough across more than one world to make them recognizable as a means of translation, through which coherence can be generated across intersecting communities.[10] These descriptions capture nicely the ambiguous position of the gold RM—it is the selected object situated at the boundary between physics and biology, between the NCI and NIST, to promote collaboration among these heterogeneous social institutions that are under pressure to work together without an obviously shared culture. The concept of boundaries is particularly important in understanding the social and technical robustness of standard objects. The gold RM is situated in a space neither belonging to scientists nor governmental officers; rather, it is a hybrid space dynamically constituted by both scientific and bureaucratic rules from different disciplines and different federal agencies. Standardization is thus a process of developing and maintaining the boundary work in settings where boundary objects are created and later on used to maintain coherence across intersecting communities.

Through boundary-object analysis, the ontology, the production, and the management of the gold RM boil down to the following aspects: (1) interpretive flexibility, (2) the correspondence of material and organizational structure, and (3) the question of scale/granularity. However, the terms *scale* and *granularity* require further clarification. The production of the gold RM cannot be understood without knowing how bureaucracy and documents function.[11] Latour and Steve Woolgar have shown that scientists are "compulsive and almost manic writers" who spend significant time making these documents: making notes on experiments, recording results in spreadsheets, drafting reports and articles, or taking up further activities that translate localized actions into circulable paper registrations.[12] The very ordinary documenting process, and the very dry paper documents surrounding scientific standardization, are as important as the very unruliness of untamed scientific objects. Nevertheless, if STS aims to deconstruct scientific authority and scientific rules, a symmetrical treatment should also be offered to explain social authority and bureaucratic rules. Only when we treat documents as analytical "ethnographic objects" and integrate them into the analysis of standard objects can the materiality of standard objects, the intimacy and tensions between the objects and human societies, be revealed.

Using the concept of the techno-bureaucratic object, the theoretical contribution of this paper is to offer a symmetrical analysis to both scientific and bureaucratic rules that produced and stabilized nanoscale standard measurement across different disciplines and different federal agencies. This study is based on one year of fieldwork in NIST and NCI, where the boundary work of creating the gold RM as a techno-bureaucratic object was performed. Both interview data and document analysis are used to capture the construction of the gold RM's social and material robustness. I analyze the technical procedures, as well as various types of bureaucratic documents generated in the process of standardization, to illustrate the construction of not only the gold RM but also the sociopolitical infrastructure that enables the conceptualization and production of the gold standard.

INTERAGENCY POLITICS IN NANOSIZE STANDARDIZATION

The FDA–NCI–NIST Memorandum of Understanding

The inconsistency of size measurement in nanodrug characterization led to an interagency collaboration among several federal units to form a size standard as a solution. To create the standard object, the NCI decided to work with NIST and the FDA, bringing metrological and regulatory concerns into nanodrug characterization. Both NIST and the FDA are active members of the NNI.

As a national laboratory responsible for the development of standards and precision measurements, NIST framed the difficulties in nanoscale measurement as "a problem of metrology"; as a regulatory agency in charge of nanodrugs approval, the FDA showed concerns that lack of a measurement standards will diminish data credibility and prohibit investigational new drug approval. As a funding agency sponsoring nanomedicine researches, the National Institute of Cancer (NCI) actively searched for models of standardizing nanodrug characterization. A government-owned, contractor operated (GOCO) facility, the Nanotechnology Characterization Laboratory (NCL), was established in 2005 under the supervision of NCI to standardize nanodrug characterization.[13] With the support of the FDA, the NCI deputy director Anna Barker and NIST director William Jeffery signed the FDA–NCI–NIST Memorandum of Understanding (MOU) in 2006 to strengthen this interagency collaboration between "life sciences for NCI and physical sciences for NIST, where the NCL brings the characterization of the biological responses while NIST brings the particle characterization to the table" (see figure 2).

The FDA–NCI–NIST MOU illustrates the work initiated by these agencies to set the "boundary" of the nanosize standard, which has to be situated in a space belonging neither to NIST nor to NCI, but to a hybrid government-owned, contractor-operated space, NCL. This setting further complicated the definition and standardization of "size," which is no longer just a metrological problem for NIST federal scientists, but also a biocompatible and regulatory question for NCL government contractors working closely with the FDA to bring nanomedicine to the market.

Contextualizing Size and Standard

What the NCI expected from NIST was a standard object with a quantified size that could be put side by side with the measured samples as a reference with which to calibrate size measurements

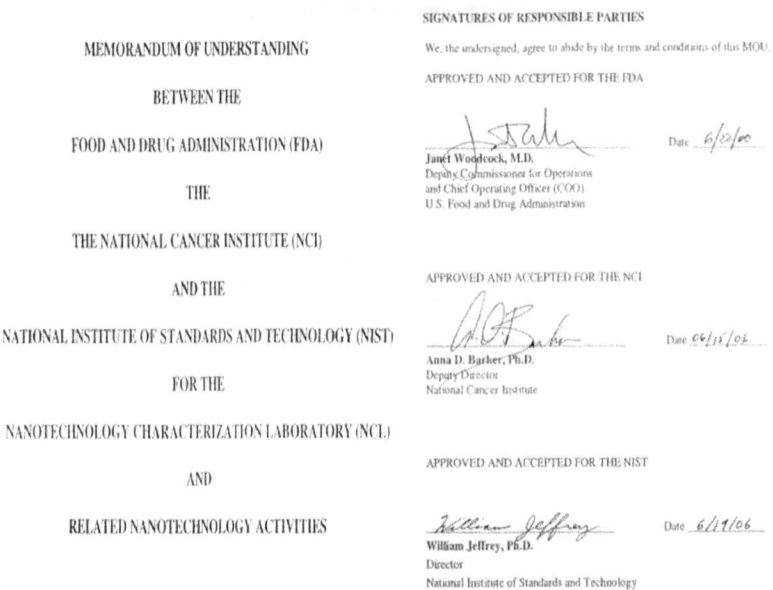

Figure 2. The Memorandum of Understanding among NCI, NIST, and the FDA.

produced at different research sites. NCI believed that making this one-dimensional size standard was an easy and straightforward task for NIST metrologists who have the expertise and tools to produce much more sophisticated standards. Nevertheless, this optimistic thinking was overwhelmingly discouraged by NIST. At an NCI–FDA–NIST joint nanomedicine symposium, Eric Steel, acting director of the program office, indirectly expressed NIST confusion in collaborating with the NCI biologists to produce size measurement: "We certainly can measure the particle size, but what does size actually mean—is it the average diameter of all particles or the actual size of an individual particle? The size distribution, shape, surfaces, crystalline form come into play, too, given that such materials are rarely perfectly defined and of symmetrical structure . . . if we don't define size properly before we measure it, we still know nothing about it."[14] Being an institute that has historical roots in the establishment of national standard weights, measures, and other industrial-based material standards, NIST customarily sees measurements as chemical and physical in nature. Angela Hight Walker, the NIST program officer who was in charge of the interagency collaboration, emphasized how standardization should be grounded in this physical tradition: "Physics has a history of changing medical science. All the big breakthroughs in medicine were physics breakthroughs, such as X-ray, imaging.... NIST has a long history in developing standards for physical measurements, and we have two Nobel Prize winners from our Physics Division."[15] What NIST argued for is more than just how to measure, but also the boundary work of redefining who has the knowledge to perform the size measurement. The disciplinary and institutional boundary between NIST physicists and NCI biologists ultimately led to a significant divergence on standardization: the NCI biologists requested a standard object with a known size value, yet NIST metrologists argued that "for NCI, size is just one number, but for NIST, a standard size claim requires thousands of measurements, statistical calibration and uncertainty analysis."[16] While the former demanded the size standard to function in biological systems, the latter questioned the feasibility of making a quantified size standard in a biological environment that is full of uncontrollable variables: "It is hard to run experiments to verify measurement techniques in biological environments, because of its lack of stability and difficulties of gathering sufficient population data. Biological systems are also [easily] perturbed from equilibrium, which causes the problem of determining whether a given measurement is real or from artifacts."[17] Anil Patri, deputy director of NCL who represented the NCI in this collaboration, disagreed with this view. He argued that "every measurement in nanomedicine is biologically oriented," even the term *stability* has to be defined in a biological context: "Drug developers care about stability. They have to be sure that the drug is stable until it reaches the tumor. After that if it is too stable, it is not releasing the drug, it's not efficacious. So stability should be tested in a biological context, like at 37°C body temperature. This is what I'm thinking about. But for NIST people from the physics or chemistry background, their stability is 'let's heat it up and see what happens.' To 300°C!"[18]

The communication between the two parties gradually turned into a matter of defending NIST physics and NCI biology: Which expertise and thought style would best run good quality measurements to determine the size standard? One NIST scientist involved in early stage negotiation recalled the extreme tensions built through the size debate: "There had already been some increasingly unpleasant phone calls about why NIST wasn't responding. NIST finally sent this long thing back, which was the worst thing they [the NIST program office] could have possibly done, which said[,] 'Well, you want us to do gold nanoparticles so we need to know the sizes, the concentrations' . . . a list of about 15 questions, which really fried them completely. I'm sure they [NCL] are like 'what the heck?'"[19] Indeed, these questions confused the NCL biologists, as they thought that according to the partnership policy listed on the MOU, NIST should be responsible for answering these questions: "People outside NIST don't know what standards mean. We conduct measurement for research purposes, so we don't have that kind of thought process. NIST should help us get through. . . . Ultimately it is not the NCI's standard; it is a NIST standard with its mark on top so it's internal NIST people who have to define what it is."[20]

In fact, the NIST program office did respond to the NCI's request, proposing to use the SRM 1963, the 100 nm polystyrene particles that NIST produced for semiconductor industrial standardization, as the size standard for nanomedicine. NIST tried to convince the NCI that SRM 1963 as the standard, with superior precision as its measurement, can be traced to the highest level international standards. However, this suggestion was firmly rejected by NCL due to the fact that polystyrene is toxic to biological systems. In contrast to NIST's position that an accurate standard should be produced in isolated controllable environments, NCL insisted that a good standard should be biocompatible with the environment in which it is situated: "Particles' quality cannot just be determined in a non-bio system because when you put them into [a] biological system, everything changes. They become new particles which are not the things you measured. For people who operate their experiments in vivo, if the particles are so variable and uncontrollable in a biological system, how can it be accepted as a good bio standard?"[21]

Standard, Institute, and Institutional Identity

The prolonged debates about size standard have lasted for almost two years; three NIST liaisons resigned from the position because none of them could avoid these circular discussions that stymied the collaboration. Yet, these technical debates were about more than fighting over who might be right or wrong. The incommensurability between NIST metrology and NCI biology is in fact rooted deeply in the organizational structures of these two parties; as Hight Walker has put it: "At NIST the word 'standard' means something which requires very strict metrological definition. . . . what NCL has in mind is an assay protocol, a quantitative measurement, not a standard."[22] Hight Walker's response reflected NIST's concerns about pursuing the interagency collaboration. Before the MOU was signed, there were already internal doubts from NIST employees, questioning the impacts of partnering with the NCI. For most of NIST's researchers, what has mattered in this collaboration is not the size of nanoparticles but the size of institutes. Compared with NIST, the NCI is a much bigger, wealthier organization. Its intramural funding, which represents only 10 percent of the total NCI budget, is larger than NIST's entire budget; the number of employees in the NCI intramural program is already double that of the entire NIST. In terms of institutional culture, unlike NCI, which has an extramural program to administer cross-institutional collaborations, NIST historically functioned as a self-sustained institute where most of the external collaborations were formed at the level of individuals rather than at the institutional level. The partnership-building policy with NCI threatened such conventions, stimulating doubts and insecurity among employees. In the 2005 internal survey, worries about benefits to NIST in the interagency collaborations were expressed: "I admit I'm not sure about this, but is going after biotech money smart for us? NIH or other governmental institutes could easily compete against us in certain areas if we are not careful."[23]

Building a symmetric collaboration under the organizational asymmetry became the first thing NIST liaisons had to demonstrate to their NIST colleagues for mobilizing their participation in the project. "We don't want NIST to look poor begging for money, and we don't want the term 'standard' which is core to our job to be misused."[24] Hight Walker guarded NIST's position as an expert of measurement, and clarified that NIST's responsibility in this collaboration was to provide "quantitative measurement" instead of providing "standards" for the NCI. As she argued for what NIST offered as "the science of measurement," as opposed to the mundane, routine, and ill-defined standards that NCI had in mind: "That [standardization] was never what I sold. Never. NIST sells science. Only 10% of NIST is about standards, our focus is on science. It had to be that we provided a type of science, a type of attitude about science that they needed, 'quantitative physical measurements' instead of 'standard.' If you read the MOU, it's mainly about physical measurements."[25]

While NIST used the term *standard* to establish its identity in this project, the NCL was also facing its identity crisis.[26] As a GOCO facility established at the NCI-Frederick, the NCL is overseen by the government agency, the NCI, and a private company, SAIC-Frederick, contracted under the Federally Funded Research and Development Center.[27] Its key mission, according to the contract, is to provide "standard assay cascade," assisting drug developers transforming their test-tube materials into FDA-approved medical products. That said, the standard measurement of nanoparticles in the drug-development context has to include physiochemical, in vitro, and in vivo characterization. As a federal contractor, the NCL has to stick with this business plan, producing results that would meet the demands of its boss, the NCI, its partner, the FDA, and its clients, the profit-oriented drug developers. Yet, being a nonfederal employee working in this highly risky field with no permanent job security, the NCL had to be strategic and practical about defining the objectives of standardization, "not to develop a standard with an infinite decimal accuracy, but 'standards' to assist our clients to get their nanoparticles into clinical trials."

The incommensurability of the NIST and NCL size-measurement proposals indicates that the size standard, rather than being viewed as a harmonizing object, in fact operates like a feedback amplifier that multiplies the degree of disagreement among scientists from these two standard institutes. The more they explored the idea of the size standard the more complicated the notion became, and the more debates were generated during their interactions. NIST and NCL scientists were not simply bargaining over the scientific definition of a size standard; they also mobilized this technical debate to draw boundaries between "us and them." The notion of "standard" is thus not an impersonal technical concept but a tool loaded with expectations, concerns, and worries regarding employee job security and institutional identity. The measurement of size, the meaning of standard, and the identity of the institutes, form mutually defined relationships and contribute to what I call the techno-bureaucratic object.

MAKING THE GOLD STANDARD

Reactivating the NIST Bureaucracy

As a consequences, the nanosize standard, an object initially associated with absolute atomic precision, turned out to be the most political tool mobilized by the agencies to protect their interest and autonomy in partnership building. The prolonged negotiation between the NCL-contracted biologists and NIST government metrologists further complicates, as M. Norton Wise argues, that "precision comes no more easily than centralized government."[28] The NCI delegates the ad hoc nanodrug standardization to a GOCO facility, NCL, which is not an intramural federal agency. Though NIST is a national standards institute, the standards it produces, according to the United States Standards Strategy, should speak for the "voluntary consensus" of standard users' community (in this case, NCL) rather than producers' mandates.[29] How to establish an effective "bureau" to coordinate and execute the metrological and pharmaceutical aspects of standardization is the key challenge of producing the gold RM.

One NIST scientist who participated in the early-stage NIST–NCI interaction described the partnership as "an arranged marriage": "Partnership has become a right thing agencies have to do for nanotechnology. It is easy for the higher level managers to say that we do physical characterization, they do biological characterization, and there's this great match . . . they signed the MOU but left the problems for bench scientists who have to actually make the relationship work . . . and we are asked to deliver a baby [the size standard] even before we barely know each other!"[30] These two agencies needed a matchmaker to repair the broken relationship between (as they might have seen one another) the stubborn NIST metrologists who always want to be in control, and the NCL cancer biologists who believe that the varieties and uncertainties in the real world cannot be simply depicted by rigid metrology.

Debby Kaiser, the chief of NIST's ceramic division, was appointed to do the job. Unlike previous liaisons who came from the physical laboratory and tended to define standard as fundamental metrology, Kaiser is a chemical engineer who identifies herself as a standard developer instead of metrologist. Her experience working with the ceramic industry gave her the insight that, before pursuing any technical collaboration, the relationship between NIST and NCI needed to be redefined: "NIST had to treat the NCI like a client, and we work with our client to produce standards based on clients' interests and the intended use of the standard." Kaiser started her reformation by redefining the NIST–NCI–FDA partnership from a political alliance into a customer relationship, to avoid competition.

Under the customer relationship, all requests from the NCI became customer demands that needed to be satisfied. Kaiser thus agreed to use the biocompatible gold nanoparticle the NCI preferred as the starting material for the size standard. In addition, she also took on the concern that this standard needed to be able to characterize multiple instruments used in nanodrug characterization, as the FDA requires the size measurement to be as thorough as possible. These adjustments indicate that Kaiser needed to organize a team across different NIST operational units so that the gold nanoparticles could be measured by different instruments registered at various divisions; at the same time, she had to mobilize the SRM flowchart into the gold nanoparticle standardization by reactivating the NIST bureaucracy into an alliance.

This was a thorny problem. NIST's more than one-hundred-year history as the national physical laboratory established a strong inertia to maintain a rigid internal division of labor. There are at least three types of measurement scientists in NIST: traditional metrologists who address fundamental measurement problems in weight or length measurements, research-oriented scientists who work on pioneering measurement projects, and standards producers who work closely with industry to develop customer-oriented standards and measurement techniques.[31]

It is this internal culture that complicates cross-institutional collaboration. While negotiating with NCI about the responsibility for conducting measurements, the NIST program office was facing challenges of internal organization—traditional metrologists might think the core work NIST should do involves fundamental measurements rather than producing or refining commercial standards. Even within industry-oriented standard producers, disagreements happened due to the fact that different laboratories might receive different amounts of funding, attention, or requests, all that certainly affected their judgment about whether or not NIST should pursue the partnership. Every NIST liaison pointed out the significant problem of internal integration in partnership-building: "I spent most of my time inside NIST dealing with interactions between laboratories, trying to get enough knowledge about what 'NIST interests' are, so I can defend it when negotiating with external parties. But different divisions have their way to do things. The hardest thing is to cross our internal barriers among segmented laboratories."[32] In addition, NIST external collaborations with industries or through internal scientific research and technical services (STRS) funding policy are all laboratory-based, which inevitably cultivated a competitive, rather than collaborative, culture among the seven laboratories:

> Everyone is so concerned with funding that they just can't take the time to see who else might have a good idea to solve the technical problems. . . . Thus there's not a lot of collaboration. Certainly we have projects which involve 3 or 4 OUs [operational units] such as the Competence Projects that formally get funded at the NIST Director's level. What I found in practice is that everybody goes their own way, does their own little piece. "Give me the money; I want to do what I want to do with the money." I try to be open but there're still lots of people who get upset.[33]

The self-absorbed institutional culture and internal bureaucratic rigidity challenged Kaiser's plan of forming a multi-OUs team. An episode occurred when she tried to recruit researchers from different laboratories that possessed instrumentation and expertise in electron microscope imaging to work on the gold nanoparticles characterization that the NCL urgently requested. The heavy load of measurement procedures this request called for, demanding repetitive experiments and detailed checks on the results, scared the NIST bureaucrats who were used to the "mind-your-own-business" bureaucratic culture: "It was a complicated project that involved a lot of techniques and people, but everyone has their own internal thing. People simply walk into their labs and gradually shy away from anything that requires a commitment, where you're accountable for something. People said 'we don't want to measure 100 samples. Forget it.'" The gold standard requires the NIST bureaucracy to undertake the measurement steps indicated on the SRM flowchart; however, the existing bureaucratic structure has not been activated to help achieve this goal. Being an experienced manager, Kaiser understood that standard development is not just about measurement, but a problem of management. A new managerial plan has to be created to redirect the flow of resources from a single laboratory to multiple OUs, changing the staff's inveterate thought pattern to focus simply on bureaucratic routines.

From Measurement to Management

Though acting as a liaison, Kaiser did not have an official position coordinating the research activities within NIST, since the collaboration with NCL was not considered a formal or routine job that NIST employees were required to do. Therefore, she needed researchers' voluntary participation. She remembered her first impression walking into NIST eighteen years earlier, when the interactions between different divisions were much more frequent and informative:

> NIST used to have something like this—the Experts' Database. When I came to NIST about 18 years ago, the first thing I did was go on the Experts' Database to know what I wanted to work on and whom to work with. You just walked down the hallway and found someone whose work interested you and talked to him. And

people tend to be very receptive with that. It's different talking with someone and saying "I need you to do these measurements on my sample because I can't measure everything I need to know."[34]

Following the same logic, she developed a customized infrastructure called Expert on the Spreadsheet, an Excel file shown in table 1. The key feature of this spreadsheet is that there is no information about the divisions, disciplines, or laboratories that each piece of equipment or individual belongs to. Instead, it uses the names of nanoparticles and the techniques suggested in NCL's assay cascade as a new category to reclassify experts and expertise within NIST. Kaiser expected this new classification system to help avoid the previous debates over whether a size reference particle should be "biological" or "physical": "We avoid the language which defines particles as "biological" or "physical"; instead we call them "particles with DLS size 30nm" or "particles with TEM size 10nm." We also don't want to remind people that oh, you are you are from MEL [the manufacturing engineering lab], and I am from MSEL [the material science and engineering lab]."[35]

She sent out the spreadsheet to the division chiefs with whom she had previously worked, asking them to fill in the required information, and gradually expanded this personal network to others who were interested in the research subjects. Active participation and a substantial interface to enroll the required expertise were what Kaiser tried to create: "This is no longer a chatty meeting; instead, they were coming in here with all their big guns, and so that first meeting was really good for that reason because it was everyone around that table. It established credibility that we were serious about this."[36]

A team constituted by cross-OU researchers was finally established. Kaiser imposed a series of managerial plans to nurture and oversee the team dynamics: NCL–NIST joint meetings were regularly held to monitor experimental progress and budget use; meeting records, progress reports, and proprietary information

Table 1. Expert on the Spreadsheet. Source: NIST nanotechnology standard working group, provided courtesy of Debra Kaiser.

NIST Measurement Capabilities and Expertise Relevant to NCL Collaboration on Nanoparticles
May 19, 2006   DRA[FT]

Material or Property	More Specifics	Existing Expertise or Existing Relevant Measurement Method	Measurement Capability*	NIST contact(s)
dendrimers	undoped, doped	general: size; dopant location		Eric Amis, 854
quantum dots	semiconductors	optical properties		Jeeseong Hwang, 844
		optical properties		Vytas Reipa, 831
lipids	liposomes	Liposome characterization, i.e., size shape, distribution	Multiangle light scattering with FFF; ultra-high resolution Bright Field microscopy	Laurie Locascio, 839
		general: size, functionality	scattering, cryo-TEM	Amis (Hudson) 854
		optical properties	vibrational spectra Raman	Angie Hight Walker, 844
silicon-based	silicon oxide or nitride	physical characterization		Debbie Kaiser, 852
	nanoporous silicon	particle counting, optical properties		Lili Wang, 831
metals	e.g., gold	physical characterization		Frank Gayle, 855
inorganic	e.g., iron oxide	physical characterization		Debbie Kaiser, 852
crystalline carbon	e.g., fullerenes	optical properties	Raman	Angie Hight Walker, 844
size	hydrodynamic size	dynamic light scattering (DLS)	0.5 to 1000 nm; all materials	Vince Hackley, 852
		dynamic light scattering (DLS)	liposomes (0.5-1000 nm)	Amis (Chastek) 854
	in solution	small-angle x-ray scattering (SAXS)	1 to 200 nm; also morphology	Amis (Bauer, Jones, Vog) 854
			liposomes (0.5-1000 nm)	Amis (Bauer, Jones, Vog) 854
			dendrimers (Compare lab SAXS with synchrotron results)	Amis (Bauer, Jones, Vog) 854
		ultra small-angle x-ray scattering	1 to 200 nm; also morphology	Andrew Allen, 852
		small angle neutron scattering (SANS)	1 to 1000 nm; also morphology	Andrew Allen, 852
			1 to 1000 nm; also morphology	Amis (Bauer, Jones, Vog) 854
				Andrew Allen, 852
			liposomes (0.5-1000 nm)	Amis (Bauer, Jones, Vog) 854
		laser diffraction	100 nm to 100 μm	Clarissa Ferraris, 861
		x-ray disc centrifuge	10 to 1000 nm; inorganic mainly	Jim Kelly, 852
		acoustic attenuation spectrometry	2 to 2000 nm; concentrated	Vince Hackley, 852

were carefully controlled, where "all hard copies of the proposals received to date should be destroyed, all written material concerning the NCL nanoparticles must be reviewed and approved by the NIST project coordinator prior to being presented at internal or external meetings or conferences."[37] In addition, testing samples and measurement data were strictly limited to those assigned for the job. These close monitoring strategies consisted of a chain of commands to keep people from different divisions on the same track, effectively reducing the instability embedded in the cross-OU and interagency collaboration.

The spreadsheet serves a critical role in standardization. It acts like the Janus with two faces looking simultaneously to the future and to the past, presiding over the beginning and ending of conflict. On one side it temporarily liberates the NIST staff from their daily bureaucratic routines when working with the NCI; on the other side, it enrolls the NCI, the FDA, the nanodrug characterization, or those who, in the past, did not fit within the NIST scope into the NIST bureaucracy.[38] In other words, the spreadsheet as a bureaucratic document reactivated the NIST bureaucracy into the reference material's production. The translation between science and bureaucracy made by the spreadsheet is extremely crucial, as it illustrates that standard objects are more than precision measurements or the products of social consensus; they are techno-bureaucratic objects defined in a deliberately constructed organization and constructed by a series of scientific and bureaucratic operations. In the following discussion, I will address how the materiality of this techno-bureaucratic object is formed by the three characteristics of bureaucratic organization in Weberian sociology: bureaucratic indifference, bureaucratic rigidity, and bureaucratic mundaneness.

Bureaucratic Indifference: Harmonizing Multiple "Truths"

Since the FDA did not specify the type of instrument required for nanodrug characterization, the NCL decided to make the size standard applicable to several instruments that have been popularly used in nanoscale size measurement. Scientifically, these multiple ways of measuring particles' size, according to their different operating principles and sample preparation, give different numerical expressions for the size values. However, this means that NIST has to produce a multiple-value standard in order to calibrate multiple instruments.

Making such a standard presented a new challenge for NIST, as all size standards NIST produced in the past were of a single value targeted to a specific sort of instrument's calibration. Kaiser decided to list all the nonreconciled values each measurement

Table 2. Different size values obtained from multiple instruments. Source: Report of Investigation of Gold RM, NIST SRM Office.

Technique	Analyte Form	Particle Size (nm)
Atomic Force Microscopy	dry, deposited on substrate	8.5 ± 0.3
Scanning Electron Microscopy	dry, deposited on substrate	9.9 ± 0.1
Transmission Electron Microscopy	dry, deposited on substrate	8.9 ± 0.1
Differential Mobility Analysis	dry, aerosol	11.3 ± 0.1
Dynamic Light Scattering	liquid suspension	13.5 ± 0.1
Small-Angle X-ray Scattering	liquid suspension	9.1 ± 1.8

technique generated: hydrodynamic size given by DLS; dry size given by AFM, TEM, and SEM; and aerosol size given by differential mobility analyses (see table 2): "The interesting thing about the gold RM is you think you will get one answer, but there is no one answer. We learned this every time we had a meeting about how to get all these techniques to converge, and realized that they're not going to. It's not that they're inconsistent; they're measuring different things about the particle. In fact, we'll start to worry if they all come up with the same value."[39]

NIST's decision not to report the size of the gold RM in a single value format induced doubts within the metrological community. A German dimensional metrologist from Physikalisch-Technische Bundesanstalt expressed his disagreement about using multiple-size measurement during his visit to NIST, questioning whether NIST can ever claim that the measurement error is 0.1 while the error bars among three techniques are larger than 1 nm.[40] Multiple values, from a traditional metrological point of view, are inconsistent with the notion of one true value in traditional metrology. It further caused confusion about which measurement is more accurate and which number is closer to the true value of the particles' size.

"We don't want to reconcile anything, this is a multiple-size RM," Kaiser insisted. In order to avoid controversies among different measurements, which could jeopardize the credibility of the standard, Kaiser mobilized the NIST bureaucratic structure for data collection. Measurement statistics offer one example here. Producing statistical results for a multisize standard was actually a "scientific research question" that NIST statisticians had not encountered before and their interest was piqued. But addressing this question was not encouraged since it would delay the release of the RM and be a distraction from efficient completion of the project:

> This [multiple size reporting] is new for the Statistical Engineering Division. Normally they get data from one technique and come up with an expectation of what the mean value is, and what the uncertainty about the mean value is. Now we're asking them to do something far more complicated: we're giving them a series of distributions produced with different techniques. How the distributions relate to each other is a new and exciting problem for them. Everybody says "what about the scientific aspect of this?" But . . . wait, let's just get the RM out first.[41]

The scientific question about the correlation among different measurement systems was avoided by the operation of the NIST bureaucracy. Researchers were only allowed to go to the group meeting specific to their assigned measurement task and reported data only to the group leader; staff from the statistical engineering division would only offer statistical results for the laboratories without questioning their research agenda. In the official report of the size measurement, seven numbers obtained from seven instruments were juxtaposed in a particular order without mentioning the correlations and comparisons among these different representations. The technical hierarchy of measurement data does not imply any inherent technical superiority about which value is more "true"; rather, it is a constructed order according to the availability of instruments to avoid potential conflicts among these different representations.

The example illustrates the mutual shaping of the social and the natural order. Whether different experimental systems were regarded as comparable, compatible, or competitive is not self-evident but depends on how the data and the labor that produced the data are managed. This information order was established through what I call *bureaucratic indifference*: each division has its own expertise and responsibility and should function under this division of labor. The bureaucratic "mind-your-own-business" culture was mobilized, under this situation, to provide the best alliance for purposes of standard-making. It offered the infrastructure to create an information order for harmonizing these "multiple truths" generated by different experimental systems, filtering out any unwanted information and local contingencies that could harm uniformity.

Bureaucratic Rigidity: The Standards' Precision

Although the organizational efforts determine a great deal of how a robust nanoscale measurement could and should be performed, bench-level administration and instrumentational choices are inseparable from institutional factors, particularly in the construction of metrologically valid "nanoscale precision." According to the International Vocabulary of Metrology, *precision* is defined as "the closeness of agreement between indications or measured quantity values obtained by replicate measurements on the same or similar objects under specified conditions." This definition highlights precision not as a static status, but as the "action" of obtaining reproducible measurements. It further indicates that to claim the gold RM as a credible standard with well-defined precision, one has to be certain that the size value of the gold RM can be consistently reproduced.[42]

Nanoscale precision, from the angle of reproducibility, becomes a problem of collective and coherent action, therefore bringing social order to the center of standardization. Harry Collins argues from a microlevel ethnographic viewpoint that reproducibility requires a "form of life" in which shared culture and tacit skills enable actors' consistent rule-following.[43] Donald MacKenzie, from a broader sociological perspective, demonstrates the historical, social, and cultural embeddedness in the invention of missile accuracy.[44] In the case of nanoscale precision, an even "harder" fact that seems to depend exclusively on atomic-scale instrumentation, I will further argue that the shared form of life has to be defined in bureaucracy, by demonstrating the correlation between precision standard, bureaucratic rigidity and routineness. I shall demonstrate this point by analyzing the indispensable role of NIST's statistical engineering division (SED) in gold RM production.

A unique characteristic of the gold RM is its existence in two contrasting contexts. In the context of commercialization, standards are considered products that need to reflect market expectation, as Bob Watters, the director of the NIST SRM office, pointed out: "Absolute precision and accuracy is meaningless when you think of an SRM as a product and users as NIST clients. If there is no market, those which cannot be popularized within the community cannot be called a standard."[45] Yet, within a metrological context, developing a standard has to follow very strict guidelines in instrument operation, error identification, and uncertainty calculation. NIST has made several adjustments to meet the NCI's demand, including abandoning the idea of using the highly precise standard SRM 1963 as the size standard. Yet, Kaiser did not view these adjustments as "compromise": "It's pretty much this [gold RM] or nothing. However, this does not imply that NIST compromised with NCL on practical concerns or ceded control over the RM's accuracy. Eventually this will be a NIST RM, and there is no compromise in the RM measurement in terms of precision and accuracy."[46] Her comment implies that "precision" for the gold RM should be considered an actors' term and a result of actors' action. To reconcile commercial demands with metrological principles in determining precision is the state of the art of making the gold standard.

In metrology, precision is dictated by "random errors" resulting from unpredictable and unknown variations in the experiment; for instance, room temperature fluctuations, mechanical vibrations, cosmic rays, etc. To improve precision means reducing the role of random errors. This can be accomplished using several strategies such as increasing the number of measurements, improving sampling procedures, using strict measurement protocols, or using more statistically efficient analytic methods. Statistics thus play a crucial role in crafting precision. Take the homogeneity study as an example. This is a step early in the development process to confirm that particle-size distribution in each and every bottle containing a substance is acceptably homogeneous and that the particle distribution remains stable. It must be done before any meaningful standard measurement is performed, otherwise the size measurement will be problematic: a measurement made using that sample in the future will not represent the average properties of the particles stated in the certificate.

The statistical knowledge and bureaucratic machinery, as Ian Hacking has nicely demonstrated, are two sides of a coin in modern forms of governance.[47] In the production of the gold RM, the SED is in charge of the homogeneity statement, which entails the design of sampling and measurement strategies, consultation on issues involving precision statements and uncertainty estimations, and the development of standardized statistical design and analysis templates used by the labs for SRM projects. From SED's viewpoint, it is statistically meaningless to take just one or two measurements to assure the homogeneity of the particles. Therefore, to validate homogeneity, the SED will require experimenters to divide a batch of particles into hundreds of bottles, and perform several measurements on a selected number of bottles to prove that the entire batch of particles is well distributed without any aggregation during the bottling process. That said, homogeneity is a mundane and time-consuming process, as same procedures have to be repeated hundreds of times without any slight modification.

Experimenters need to have extreme patience to conduct these repetitious, mundane tasks that do not produce fresh ideas or new results. "Project managers want to finish this part as soon as possible, as it is costly and could delay the whole standardization process. If they have the right to choose to do measurements 4 times instead of 400 times, there will be no hesitation," according to one of the SED staff in charge of the nanosize standard project.

The SED statisticians and their statistical modeling have played the "bad cop" to gatekeep measurement results and data interpretation. However, bureaucratic rigidity allows them to not worry about concerns of time, cost, or market demands. Their sole responsibility is to check whether the measurements that NIST laboratories provide meet statistical requirements sufficiently to claim the precision statement. Measurement advice is given according to the operation of a statistical model; scientists who are in charge of the measurement task, no matter how well-reputed or senior they are, cannot negotiate but only follow the instructions, as the gold RM project leader indicated: "Right now we probably have 400 bottles of each different size. I've been pulling them and handing them out to the other people to do the measurements. It's tedious, but there's no compromise."[48] In other words, bureaucratic rigidity offers the SED the authority to collect a large quantity of data, filtering out unwanted information and mitigating local contingencies (random errors) generated in the process of measurement. Statistics and bureaucracy go hand in hand to transform precision measurement from a heroic scientific representation into a series of routine practices.

Bureaucratic Mundaneness: The Standards' Robustness

Last but not least, bureaucratic mundaneness plays a critical role in securing the gold RM's material robustness. This might be counterintuitive for many nanotechnology innovators who think the ability to capture and stabilize scientific unknowns in the nanoworld should be grounded in advanced instrumentation. However, the operation of NIST's measurement service division (MSD) in gold RM certification and postproduction services illustrates the deep dependence between the most mundane practice and most advanced science.

As Wittgenstein suggests, what could give a word its meaning is a rule for its use. For the gold RM to be recognized as a standard, NIST needs to guarantee that it can be properly used and followed by users. In other words, the users' perspective (also mentioned by Senaga in this volume) is as equally important as a standardizer's perspective. The MSD is the NIST office in charge of the administrative aspects of and product support for RMs, to make sure that they can be properly delivered to and accurately followed by customers. Its business ranges from the metrological support such as laboratory accreditation, instrument calibration, and specialized material processing, all the way to commercial activities such as packaging, labeling, pricing, warehousing, and sales and distribution services for all RM products. In other words, MSD acts in something of a housekeeping role, dealing with all sorts of mundane tasks associated with the development and application of standards.

The degree of mundaneness associated with nanotechnology standardization is vividly illustrated by the gold RM "Report of Investigation," a document released with the RM material by the MSD to inform users of the right procedure of using the standard. The Report documents both material and social properties of the gold RM (see figure 3a), from expiration dates to the personnel involved in the standardization process. In addition, there are two sections, "Notice and Warning to Users" and "Instruction for Use," that give detailed descriptions about sample handling and storage. Even for such a trivial action as opening the bottle in which the reference specimen is packaged, there is a rule to follow. According to the instructions for use, one should open the bottle by "gently flicking the nipple with forefinger while tilting the ampoule,

National Institute of Standards & Technology
Report of Investigation

Reference Material 8011

Gold Nanoparticles, Nominal 10 nm Diameter

This Reference Material (RM) is intended primarily to evaluate and qualify methodology and/or instrument performance related to the physical dimensional characterization of nanoscale particles used in pre-clinical biomedical research. The RM may also be useful in the development and evaluation of in vitro assays designed to assess the biological response (e.g., cytotoxity, hemolysis) of nanomaterials, and for use in interlaboratory test comparisons. RM 8011 consists of nominally 5 mL of citrate-stabilized Au nanoparticles in an aqueous suspension, supplied in hermetically sealed pre-scored glass ampoules sterilized by gamma irradiation. A unit consists of two 5 mL ampoules. The suspension contains primary particles (monomers) and a small percentage of clusters of primary particles.

Expiration of Material: The reference values for RM 8011 are valid, within the measurement uncertainties specified, until 31 December 2012, provided the RM is handled in accordance with the instructions given in this report (see "Instructions for Use"). However, the size distribution may be altered and the RM invalidated if the material is contaminated or handled improperly.

Maintenance of Reference Values: NIST will monitor representative samples from this RM lot over the period of its validity. If substantive changes occur that affect the reference values before the expiration date, NIST will notify the purchaser. Registration (see attached sheet) will facilitate notification.

The overall technical coordination for material procurement, processing and measurement activities was conducted by V.A. Hackley and J.F. Kelly of the NIST Ceramics Division.

Reference and informational value measurements were performed at NIST by the following: NIST Analytical Chemistry Division: T.A. Butler, R. Case, K.W. Pratt, L.C. Sander and M.R. Winchester. NIST Ceramics Division: A.J. Allen, T.J. Cho, J. Grobelny, V.A. Hackley, D.-I. Kim and P. Namboodiri. NIST Metallurgy Division: J.E. Bonevich and A.J. Shapiro. NIST Polymers Division: M.L. Becker, D.L. Ho, A. Karim and B.M. Vogel. NIST Precision Engineering Division: B. Ming and A.E. Vladár. NIST Process Measurements Division: L.F. Pease III, M.J. Tarlov, D.H. Tsai, M.R. Zachariah and R.A. Zangmeister.

Statistical consultation on measurement design and analysis of the reference value data were performed by A.L. Aviles of the NIST Statistical Engineering Division.

Additional technical and coordination aspects were provided by the following: R.F. Cook, W.K. Haller and D.L. Kaiser of the NIST Ceramics Division.

Support aspects involved in the preparation and issuance of this RM were coordinated through the NIST Measurement Services Division.

RM 8011 was developed at the request of the National Cancer Institute (NCI). Development and production costs were subsidized by NCI.

Debra L. Kaiser, Chief
Ceramics Division

Gaithersburg, MD 20899
Report Issue Date: 13 December 2007

Robert L. Watters, Jr., Chief
Measurement Services Division

Figures 3a and 3b. "Report of Investigation of NIST Gold Nanoparticle Reference Material"; a paragraph from the "Report of Investigation," tutoring users on how to open the bottle of gold RM.

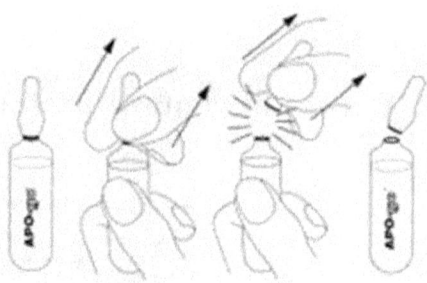

Prior to opening, the glass ampoule containing the RM should be gently inverted several times to insure homogeneity and resuspension of any settled particles. Liquid retained in the upper portion of the ampoule (the nipple), can be dislodged by gently flicking the nipple with forefinger while tilting the ampoule. The ampoule is pre-scored and should be opened by applying moderate pressure with one's thumb to snap off the nipple. It is recommended that the contents of an ampoule be used the same day as opened. Clean laboratory sealing film can be

... applying moderate pressure with one's thumb to snap off the nipple" (see figure 3b).

These mundane rules not only serve the purpose of disciplining standard users; more important, they protect the authority of the gold RM through mobilizing the NIST bureaucratic machine. According to the MSD manager, the office often received phone calls from external users questioning the quality of the standards, as they had been unable to reproduce the results claimed on the standard certificate. Yet most of the complaints ceased after MSD staff requested further information and paperwork for troubleshooting: To file an investigation, users have to report every experimental detail, making sure that every instruction listed on the certificate has been carefully followed, to validate their challenges of the NIST RMs. In other words, to question the material robustness of the RM, one needs to be prepared not only to encounter the object itself, but also to fight with the whole NIST bureaucracy.

Bureaucratic mundaneness therefore guarantees the standards' material robustness. Only through a thorough execution of these "unsurprising," "non-innovative" routines, can the gold RM earn its social and technical credibility as a trustworthy standard:

Nothing gets released until it goes through the Measurement Services Division. They can call a stop if they don't approve the documents we submit. For example if I tried to put an RM through and I didn't have all the statistics in place, they would say "Forget it. Go back and do the statistics." There are these different gates that you have to go through so NIST is not going to release anything unless every box is checked. Even as a project manager, I cannot go off by myself and throw the results to their asking for their stamps. It has many approval steps along the way to ensure that the standard is acceptable.[49]

Day-to-day deployments of materials and procedures the MSD offers are thus not something that blocks or follows innovation, making stages in any simple sense; rather, they continue the activities of invention by securing the production of quality, social interactions as well as the exercise of bureaucratic power that contribute to the material and social robustness of standards.

CONCLUSION

In January 2008, the NIST SRM office finally released the gold RMs—RM 8011, 8012, and 8013, which contain standard gold nanoparticles of 10 nm, 30 nm, and 60 nm, respectively, stored in liquid form (see figure 4). Those high-level disputes and prolonged negotiations seem all to be put into this 5 ml bottle. Yet, this dramatic scale change of turning the complex institutional politics into a neat nanosize standard depends on more than just measurement, but also on the politics of measurement. It took three and a half years for NIST and NCL to make the size standard, because the whole process is not just a production of a material object, but also many kinds of social "objects"—the FDA–NCI–NIST MOU; the letter from the NCI nanotechnology director; thousands of email exchanges and meetings between NCI and NIST staff; and experts on the spreadsheet, the "Report of Investigation"—to constitute

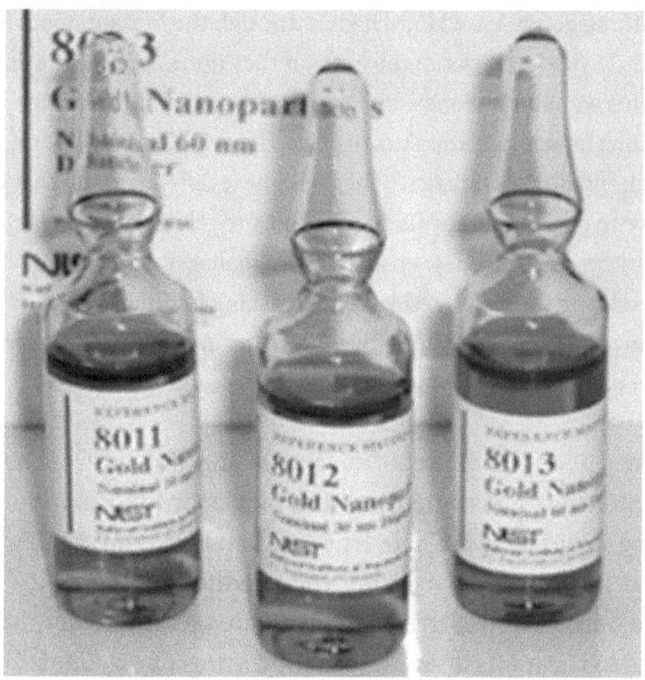

Figure 4. NIST certified-gold RMs with three different sizes—10 nm, RM 8011; 30 nm, RM 8012; and 60 nm, RM8013.

the gold nanosize standard. In other words, the bureaucratic system needs to be mobilized and carefully inserted into the RM production to "institutionalize" the gold nanoparticles, the scientifically untamed and uncertain object, into a technically, socially, and politically robust standard.

The standard-making, from this perspective, is neither a pure technical process nor a process dictated solely by social factors. It is a series of actions taking place in the techno-bureaucratic space, through which the construction of RM's precision and material structure, and the reformation of the interagency relationship and organizational structure, are coproduced. The mutual shaping of bureaucratic and material cultures brings the gold RM to life, from untamed particles into a NIST certified material associated with a series social and material documents—a well-defined size, secured packaging, "Report of Investigation," the users manual, and the NIST–NCI–FDA MOU. It is these hybrid social and technical properties that form the "new materiality" of nanosize standards as a "techno-bureaucratic object" derived from the close ties between precision and bureaucracy.

The metaphor of "the ship in a bottle" that Collins used to describe the construction of universality captures well this new materiality of a techno-bureaucratic standard object.[50] One can find that the logo of the Department of Commerce appearing on the "Report of Investigation" is a ship full of goods. It conveys a symbolic meaning off a standard as something that is expected to travel like the ship to enhance commercialization and global trade. However, for this delicate standard ship to travel well, it has to be bottled. Efforts to build the ship in a bottle bear some resemblance to the process of standardization—they are messy, unstable, full of contingencies, and demand lots of craftiness. Yet, once the ship is constructed, those "nonscientific" elements become invisible—the ship stands beautifully in the bottle, just like the gold RM stores stably in the 5 ml ampoule. Their neat existence makes us forget about their painstaking coming-into-being. This closeness and forgetfulness is state-of-art standardization.

The biographies of the gold RM offer an alternative with which to rethink the material agency of standard objects. Standards become more and more important in contemporary partnership-based research, although not because they bring in order of integration. Quite the opposite. Their power of harmonization, coordination, and discipline do not come from themselves but from the organization where they are produced, as it is the collective action that turns the unstable gold nanoparticle into the credible gold standard. In other words, standard-making is a declaration and a series of actions made in the techno-bureaucratic space. Our understanding of standard objects thus should go beyond pure social constructivism or naïve scientific realism: these objects are neither rigidly defined entities that solely come from highly precise instrumentations, nor are they powerful actants that automatically own social agency to intervene human action. Instead, these objects carry many social expectations and social agreements; at the same time, they have to be constrained with layer-by-layer scientific and bureaucratic rules to prevent any arbitrary constructions, either scientifically or socially. There is no single individual, laboratory, or institution that owns the knowledge and techniques of producing these techno-bureaucratic objects, no matter how technically superior or politically authoritative. Standard objects will start taking shape and become the spokespersons for humans only when all interested parties involved in the standardization process stop competing and start communicating and negotiating the degree of tolerance to deal with differences among them.

Finally, the story of the gold RM shows that the concepts of size and precision used in legitimating the development of nanotechnology (such as "size matters" or "precision dominates") do not engage with reality where size and precision are not determinants but battlefields of nanotechnology. The problematic usages of these terms have caused, and are continuing to allow, unrealistic policy statements and questionable decision making to guide current nanotechnology investment and regulatory policies. Many regulatory debates are still centered on "size" and size quantification, which inevitably limits the solution solely to developing precision measurements and accumulating more scientific data. The biographies of the gold RMs offer the benefit of the doubt to this science-driven policy-making; Richard Feynman might be surprised that his 1959 speech at Caltech, "There Is Plenty of Room at the Bottom," made him the spokesperson of nanotechnology.[51] Yet, he forgot to mention that to claim room in the natural world, the room in the social world needs to be sorted out. This social space indicates that nanoscale precision and standardization are not just science underway in an atomic world but politics in our public life, about expertise, government, and governance. There's indeed plenty of room at the bottom for more transparent social negotiations and dialogues for a more down-to-earth nanotechnology policy.

NOTES

1. National Nanotechnology Initiative, "What It Is and How It Works," accessed August 4, 2015, https://www.nano.gov/nanotech-101/what nano.gov. An example of the canonical history of nanotechnology can be found at "History of Nanotechnology," accessed January 20, 2020, http://www.trynano.org/about/history-nanotechnology.
2. The definition was provided in his 1974 paper "On the Basic Concept of 'Nano-Technology,'" part 2, Proceedings of the International Conference Production Engineering, Japan Society of Precision Engineering, Tokyo, Japan. The European Society for Precision Engineering and Nanotechnology presented Professor Taniguchi with its first Lifetime Achievement Award in Bremen, in May 1999, with the inscription: "In recognition of his unique and outstanding contributions to research and development in the ultra precision materials processing technologies and in 1974, being the first to formulate and use the term Nanotechnology. Through his vision, writings and example of total dedication to his field of endeavour he has stimulated the development of what will be one of the dominant technologies of the 21st Century." For more information, see N. Taniguchi, "On the Basic Concept of Nanotechnology," Proceedings of the International Conference on Production Engineering, 1974, 18–23.
3. See Interagency Working Group on Nanoscience, Engineering and Technology, "Nanotechnology: Shaping the World Atom by Atom," National Science and Technology Council, Committee on Technology, Washington, DC, September 1999. For the full report, see http://www.wtec.org/loyola/nano/IWGN.Public.Brochure/IWGN.Nanotechnology.Brochure.pdf.
4. Simon Schaffer, "Late Victorian Metrology and Its Instrumentation: A Manufactory of Ohms," in *Invisible Connections: Instruments, Institutions and Science*, eds. Robert Bud and Susan Cozzens (Bellingham, WA: SPIE Optical Engineering Press, 1992), 23–56; Ken Alder, *The Measure of All Things: The Seven-Year Odyssey and Hidden Error that Transformed the World* (New York: The Free Press, 2002), 137.
5. Schaffer, S. "Late Victorian Metrology"; Amy Slaton, *Reinforced Concrete and the Modernization of American Building, 1900–1930* (Baltimore, MD: Johns

Hopkins University Press, 2001); Joseph O'Connell, "Metrology: The Creation of Universality by the Circulation of Particulars," *Social Studies of Science* 23, no. 1 (1993): 129–73; Stefan Timmermans and Marc Berg, "Standardization in Action: Achieving Local Universality through Medical Protocols," *Social Studies of Science* 27, no. 2 (1997): 298.

6. In fact, Porter made the following comment about Latour's interpretation of standardization: "There is something deeply right about Bruno Latour's phrase 'center of calculation' to describe the point from which empires are administrated and about his emphasis on this in a book about technology and science"; Theodore M. Porter, *Trust in Numbers* (Princeton, NJ: Princeton University Press, 1995), 224.

7. On the theme of "center of calculation," see Bruno Latour, *Science in Action: How to Follow Scientists and Engineers through Society* (Cambridge, MA: Harvard University Press, 1987), 216–57. He mentions that the US government spends 6 percent of its gross national product, three times that of the scientific research budget, on maintaining physical constants.

8. Gerald M. Britan and Ronald Cohen, eds., *Hierarchy and Society: Anthropological Perspectives on Bureaucracy* (Philadelphia, PA: Institute for the Study of Human Issues, 1980).

9. Cyrus Mody and Michael Lynch, "Test Objects and Other Epistemic Things: A History of a Nanoscale Object," *British Journal for the History of Science* 43, no. 3 (2010): 423–58.

10. Susan Leigh Star and James Griesemer, "Institutional Ecology, 'Translations,' and Boundary Objects: Amateurs and Professionals in Berkeley's Museum of Vertebrate Zoology, 1907–39," *Social Studies of Science* 19, no. 3 (1989): 387–420.

11. Founded in 1901, NIST was originally rnamed the National Bureau of Standards (NBS) and in charge of US weights and measures under the Department of Commerce. It became the National Institute of Standards and Technology, or NIST, in 1988.

12. Bruno Latour and Steve Woolgar, *Laboratory Life: The Construction of Scientific Facts* (Princeton, NJ: Princeton University Press, 1986); Bruno Latour, "Drawing Things Together," in *Representation in Scientific Practice*, eds. Michael Lynch and Steve Woolgar (Cambridge, MA: MIT Press, 1990), 54.

13. A facility owned by a federal agency but operated in whole or part by contractor(s). This is a unique operational model developed by the US government during World War II to facilitate the production of scientific research and technological products for national needs. A GOCO facility can be managed by a university, nonprofit, or for-profit organization, depending on its

mission as defined by the government. The NCI GOCO is operated by a private company, SAIC-Frederick. The NCL is part of it.

14. The comment was made in his presentation "Bionanotechnology: The Role of Measurements & Standards," NCI–FDA–NIST Nanomedicine Symposium, Bethesda, MD, May 19, 2005.
15. Angela Hight Walker, personal communication with the author, June 14, 2006.
16. Walker, personal communication.
17. Eric Steel, "Bionanotechnology," NNI Workshop on Instrumentation & Metrology for Nanotechnology, Gaithersburg, MD, January 27–29, 2004.
18. Anil Patri, personal communication with the author, April 24, 2006.
19. Debra Kaiser, personal communication with the author, May 15, 2007.
20. Patri, personal communication with the author, April 24, 2006.
21. Anil Patri, personal communication with the author, May 10, 2006.
22. Angela Hight Walker, personal communication with the author, June 27, 2006.
23. Walker, personal communication.
24. Walker, personal communication.
25. Walker, personal communication.
26. The social history of the NCL can be found at Sharon Ku, "Forming Nanobio Expertise: One Organization's Journey on the Road to Translational Nanomedicine," *Nanomedicine and Nanotechnology* 4 (2012): 366–77.
27. First established during World War II, the Federally Funded Research and Development Center (FFRDC) was created to assist federal agencies engaged in scientific research, systems development, and systems acquisition in executing their missions in support of national objectives as efficiently and quickly as possible. An FFRDC can be operated using four mechanisms: GOGO (government-owned, government operated), GOCO (government-owned, contractor-operated), COCO (contractor-owned, contractor-operated), or GLGO (government-leased, government-operated), depending on its mission. NCI-Frederick is the only GOGO facility under the Department of Health and Human Services. It was established in the 1970s at a time when "fighting the war on cancer" was portrayed as a national mission.
28. M. Norton Wise, *The Values of Precision* (Princeton, NJ: Princeton University Press, 1995), 93.
29. US Congress and the executive branch have recognized the benefits of voluntary consensus standards, through the National Technology Transfer and Advancement Act of 1995. Present US policies for government production and procurement of standards—as articulated in ANSI's US National

Standards Strategy and OMB Circular A-119 "Federal Participation in the Development and Use of Voluntary Consensus Standards and in Conformity Assessment Activities" (1998)—emphasize the virtues of the voluntary consensus-standardization process. According to the definition given by Circular A-119, voluntary consensus standards are standards developed or adopted by voluntary consensus standards bodies, both domestic and international. These standards include provisions requiring that owners of relevant intellectual property have agreed to make that intellectual property available on a nondiscriminatory, royalty-free, or reasonable royalty basis to all interested parties. For more information, see http://www.whitehouse.gov/omb/circulars_a119; for reference, see Roger B. Marks and Robert E. Hebner, "Government/Industry Interactions in the Global Standards System," in *The Standards Edge: Dynamic Tension*, ed. Sherrie Bolin (Ann Arbor, MI: Sheridan Books, 2004), 103–14.

30. Personal communication with an anonymous NIST scientist during my fieldwork at NIST in the summer of 2006.
31. Four researchers at NIST have been awarded Nobel Prizes in physics for their work in advancing the atomic clock based on quantum physics: William D. Phillips in 1997, Eric A. Cornell in 2001, John L. Hall in 2005, and David J. Wineland in 2012. It is the largest number for any US government laboratory; the work of these four physicists represents atomic precision as a state-of-the-art matter, which the institute views as a core value of their measurement expertise.
32. Angela Hight Walker, personal communication, April 15, 2006.
33. Debra Kaiser, personal communication, May 10, 2007.
34. Kaiser, personal communication.
35. Kaiser, personal communication.
36. Kaiser, personal communication.
37. Internal document circulated among members of the gold RM team.
38. The Janus metaphor indeed nicely explains the sociological meaning of standardization. Any standard should not be viewed simply as a fixed rule to be followed; quite the opposite, standards require actors' active interpretation, declaration, and demonstration.
39. Debra Kaiser, personal communication with the author, May 10, 2007.
40. The information was obtained from a NIST research scientist, John Dagata, through personal communication with the author, July 31, 2007, during fieldwork.
41. Robert Cook, assistant manager of the gold RM project, personal communication with the author, July 27, 2007.

42. Strictly speaking, our common usage of the term *precision* contains two notions: precision and accuracy from a metrological perspective. According to VIM3, *accuracy* is defined as "closeness of agreement between a measurand quantity value and a true quantity value of a measurand." Unlike precision, which is a factor associated with the occurrence of random errors, accuracy is associated with systematic errors such as bad instrument calibration. Accuracy thus cannot be treated by statistics. In this chapter, I only discuss the process of crafting precision for the purpose of arguing the role of statistics and bureaucracy in standard-making. See the Joint Committee for Guides in Metrology, "International Vocabulary of Metrology-Basic and General Concepts and Associated Terms" VIM, 3rd ed. (Joint Committee for Guides in Metrology, 2008), 22.
43. Harry Collins, *Changing Order: Replication and Indution in Scientific Practice* (Chicago, IL: University of Chicago Press, [1985] 1992).
44. Donald MacKenzie, *Inventing Accuracy: A Historical Sociology of Nuclear Missile Guidance* (Cambridge, MA: MIT Press, 1990). He gave a historical explanation about why maximum accuracy became the priority concern for missile-guidance systems and showed how accuracy was "socially constructed" by different calculation systems and different organizational cultures.
45. Bob Walters, personal communication with the author, May 22, 2007.
46. Debra Kaiser, personal communication with the author, May 10, 2007.
47. Ian Hacking, "How Should We Do the History of Statistics?" In *The Foucault Effect*, ed. G. Burchill, C. Gordon, and P. Miller (Chicago, IL: University of Chicago Press, 1991), 181–95.
48. Vince Hackley, technical director of the gold RM project, personal communication with the author, June 20, 2007.
49. Debra Kaiser, personal communication with the author, September 19, 2007.
50. Collins, *Changing Order*.
51. See note 1, above.

PART II

MATERIALS PRODUCED, LABOR DIRECTED

CHAPTER FOUR

THE SCIENTIFIC CO-OP

Cloning Oranges and Democracy in the Progressive Era

Tiago Saraiva

Well-tended landscapes epitomize good government. There is no more eloquent depiction of that aphorism than Ambrogio Lorenzetti's fresco on the walls of Siena's public palace, painted in the late fourteenth century. This allegory of good and bad government, after being for many years a favorite object for political-theory scholars obsessed with its intricate symbolism and the complicated genealogy of its multiple emblems, has now become, somehow unexpectedly, part of the science and technology studies (STS) repertoire, following Bruno Latour's discussion of his "parliament of things."[1] Latour urges his readers to mimic the attitude of contemporary tourists in Tuscany, ignore the erudite tradition of Western political thought, and focus instead on the easily interpretable dual ecology painted by Lorenzetti: ruined cities, fires, barren land, war—bad government; crops, animals, farmers, arts, commerce—good government. The fresco calls attention to the actual issues that government should be concerned with in contrast to abstract theories of representation that make so much political science of dubious relevance. We are advised to put the things that populate Lorenzetti's landscapes at the core of politics instead of taking them as mere scenery accommodating the general principles of bad and good. As an alternative to the traditional parliament composed of allegedly disinterested members through whose dialogue general interest is established, we need, Latour asserts, a hybrid fora gathering humans and nonhumans alike, a true parliament of things.[2]

As cryptic and unsettling as such language may be for those not initiated in STS jargon, Latour explicitly acknowledges that he is building his proposals on the very respectable shoulders of American pragmatist philosophers; namely, John Dewey's.[3] Latour's alternative parliament draws on Dewey's attempts to overcome object/subject dichotomies, and to make pragmata, the Greek word for *things*, the basic components of a democratic society. Dewey's assemblies gathering concerned citizens to discuss the issues—or things—affecting them may be seen as the Progressive Era equivalent of the parliament of things. And Latour also follows Dewey in ascribing to science a central role in the existence of such assemblies. More than scientific experts producing matters of fact in order for politicians to make informed rational decisions, they bring new matters of concern to the forefront, offering, in addition, a method to deal with them.[4] The communitarian experimentalism that, according to Dewey, characterizes scientific undertakings should also be the rule of democratic societies.[5]

Latour's reading of Dewey tends to ignore the particular historical context of the latter. This is characteristic of STS scholarship, a field not always aware of the historical bagagge of the concepts it creatively uses.[6] As Projit Mukharji reminds us in his afterword to this volume, the importance of such historical reflexivity is only more salient in political moments such as ours in which unreflected uses of history feed authoritarian populisms. This chapter, while reproducing Latour's gesture toward Dewey, fully places the American philosopher in the Progressive Era in order to explore how new technoscientific things such as Californian oranges enabled the production of new political collectives. Historicizing Dewey not only reveals the democratic vistas opened up by the practices of cloning oranges, but it stresses as well the race-making involved in such practices. More bluntly, cloning oranges became a practice for the reproduction of whites, the exclusive members of a new, enlarged democratic collective.

It is certainly no coincidence that the care for the landscape was also central to Dewey's pedagogical proposals, a field that would make much of his fame outside academia. He was namely an enthusiast of nature study, promoting object-lessons and experience-based learning in place of books and abstract principles as main tools to cultivate would-be citizens.[7] Citizens would come into being through hands-on experiences, not the top-down transmission of scientific ideas or philosophical theories. The materialization of such program was to be found in the garden of his University of Chicago Laboratory School, a place where students became familiar with science as they cultivated their communal plot of land.[8] As plants were trained by using the most up-to- date science of the day, so students were being trained as good democrats, solving issues through the use of the scientific method. And as in Lorenzetti's fresco, the well-tended garden of the laboratory school embodied the virtuous democratic society envisioned by Dewey.

This chapter takes the citrus orchards landscape of Southern California in the early decades of the twentieth century as materializing Deweyean virtuous democratic-scientific communities. The narrative explores the development of citrus growers' cooperatives through technoscientific practices in the Los Angeles region. It argues that each new experimental system developed by scientists of the United States Department of Agriculture (USDA) and of the Citrus Experiment Station of the University of California (UC) in Riverside—systems aimed at securing the quality and quantity of the orange as commodity, as a material to be made standard for markets—contributed to the forming and strengthening of social ties among orange growers.[9] The text details the importance of cloning and phytopathology practices to the forming of a cooperative of citrus growers, a scientific co-op. In more general terms, it takes experiments with oranges as experiments in US democracy.

The social reformist character of this experience, rather than any habituated commitments to progress or efficiency, propelled the orange growers. Contrary to most visions of Californian agriculture as a mere extension of industrial capitalism habits into the farmer's realm, the chapter stresses the democratic dimensions involved in producing standardized oranges.[10] More than just bringing capitalism into the farm, what was at stake in the new practices developed in the California Fruit Growers Exchange—the citrus co-op founded in 1905—was the weaving of new social ties as an alternative to the unbridled capitalism of the railway barons of the Gilded Age. Scientific expertise, as multiple authors have noticed, was a major component of the proposals advanced by Progressive Era-reformers to regulate the US economy and thus avoid the cycles of boom and bust characteristic of the last three decades of the nineteenth century.[11]

As revealed by this chapter, Southern California landscapes would become exemplary of the interventions at the national scale, promoted by the Country Life Commission formed in 1909 under the leadership of Liberty Hyde Bailey.[12] The Commission, a quintessential institution of the Progressive Era, promised an uplift of farm life through the generalized application of science, counting on the presidential patronage of Theodore Roosevelt, who had claimed that rural America was "the backbone of our Nation's efficiency." The focus on citrus suggests a more complicated and interesting historical dynamic than the familiar top-down narrative: growers made use of state science (both at federal and state levels) as a resource to strengthen their new social organization—the

California citrus co-op—whose much-publicized success justified the growing presence of the state in the US countryside nationally from the Progressive Era onward.[13] The citrus co-op grew with the growth of the state.

While detailing the entangled social and scientific experimentalism occurring in Southern California in the first decades of the twentieth century, the chapter points at the racializing effects of this American experiment. As many other historians have stressed, the citrus belt cities were indeed segregated communities, effectively separating whites from Chinese, Japanese, and Mexican people.[14] The Californian citrus world was the result of attempting to reform democracy and capitalism, and as in other such attempts across the United States in the Progressive Era, new racial faultlines were produced.[15] Less noticed is how racial separation depended on the concrete technoscientific practices explored here and less on generic notions of white-anglo supremacy. Simply put, white supremacy was materialized through the (re)production of oranges, the new materials of the volume title.

BLUE MOLD

In 1904, G. Harold Powell, a horticulturalist from the USDA Bureau of Plant Industry at Washington, DC, arrived in Southern California to start an investigation on the decay occurring in citrus while in transit to eastern markets.[16] His presence was requested by Riverside citrus growers who claimed annual losses ranging from $500,000 to $1.5 million. Citrus decay damaged the reputation of Californian fruit among consumers, endangering an otherwise highly profitable business that constituted the backbone of Southern California's economy until the 1920s.

The blue-mold fungus had already been identified as the organism responsible for the decay occurring in railway cars during the long transcontinental journey, breaking down the structure of the fruit and destroying, in many cases, 25 percent of an entire shipment. The issue at stake was the multitude of reasons advanced to explain the observed susceptibility to the fungus of an increasing number of oranges. Types of soil, age of trees, and climatic conditions were all mentioned, as well as a supposed inherent weakness of the fruit during the blooming period. The growers, always suspicious of shippers' practices, also brought up handling methods during transportation. Disagreement on the reasons for decay led to mutual mistrust between the parties involved.

Powell promised to reestablish good relations among the different agents through a scientific survey of the citrus industry, unveiling the causes of the problem.[17] For the following three years, he and five fellow USDA assistants scrutinized in detail operations in the groves, packinghouses, cars in transit to the East, and in the main markets. Together with laboratory observations of the fungus, one thus found in Powell's report an informed discussion on the social composition and cultural values of the grower class, of the working habits of pickers, of the machinery equipping packinghouses, of railroad-shipping conditions, and of New York's wholesale fruit merchants. For Powell, there was no fault line between the natural and the human-made, both forming the ecology of funguses.

Upon examination of oranges in which the rot was just starting, Powell concluded that the area of decay started to form around a spot where the epidermis of the fruit had been previously injured by a cut or abrasion of some kind. It was also noticed that different types of injuries entailed different decay rates. The painstaking laboratory work involved in making a typology of orange injuries proved crucial for establishing the connections between the fungus and actual operation conditions of the industry. A case in point was the impression among growers and shippers that there was a wide difference in the inherent keeping quality of the oranges from different growing areas. The fruit from the upper San Bernardino Valley, for example, was said to have better shipping qualities than the oranges grown in the more humid regions near the coast. The observation of the average condition of oranges in the market seemed to confirm the general impression. But Powell's team's conclusions couldn't be clearer: after looking into the conditions in many areas where the fruit developed unusual amounts of decay, the trouble had been found, in every case, to be due not to location but to the specific conditions under which the fruit had been handled. Following the path of fruit injuries, it became possible for the team in every case to relate a high presence of decay with the care with which scale insects were held in control in the groves, the business capacity of the directors of a given growers' association, the ability of the packing-house foreman, the system of handling labor in the groves and packinghouses, the efficiency of the labor, the sanitary conditions in the packinghouses, or the technical characteristics of packinghouse equipment. Or, in Powell's own words: "After a careful study of the problem for three years, it is evident that the mechanical injury of the orange is partly inherent to the handling of a perishable crop, but that is primarily related to the economic and social conditions surrounding the citrus-fruit industry."[18]

To better understand such claims as reflecting a melding of scientific and social intentionalities, it is necessary to look more closely at Powell's analysis of injuries related to fruit picking. In his analysis, the precarities of commercial orange production come to represent a result of inseparable material and labor conditions, soluable only by changes that engage with both natural and human(made) elements of the enterprise. The thorough observation of some forty thousand fruits from several packinghouses and groves in different areas revealed that the most common type of injury was made by the point of the scissor clippers when cutting the oranges from the trees.[19] Many oranges were also injured by stem punctures produced when a fruit cut from the tree with a long stem was allowed to fall or roll against another fruit. One had to consider in addition the punctures and bruises from gravel and twigs in the bottom of picking boxes or cuts in the skin caused by the long fingernails of the pickers. It was found that there was a wide difference in the amount of injury in the fruit delivered by different growers to the same packinghouse, as well as great variation in the condition of the fruit in different picking boxes coming from the same grove. By surveying different pickers, individuals were found who had injured no less than 75 percent of the fruit they handled.

These were surprising results, even for growers who had given careful attention to all details of production but who did not realize the importance of careful picking. According to Powell, the payment of labor by the box had developed a class of pickers whose chief aim was to fill up the largest number of boxes in the shortest period of time possible. With a hoop fastened in the top of the picking sack to hold the mouth open, these fast pickers needed no more than a quick movement of the clippers to cut the oranges and shoot them directly into the bag. Such practice led to the use of sharp-pointed clippers, or shears, which the picker could handle with great facility but that were also responsible for damaging the fruit. To remediate the problem, Powell didn't limit himself to suggest the use of clippers with blunt points. An entire reorganization of the labor system was needed. First, payment was to be done by the day and not by quantity. Second, gangs of laborers under the control of packinghouse management should undertake the picking of the fruit. Foremen appointed by the packinghouse were to oversee handling operations, controlling the gestures of pickers who now had their nails regularly inspected to avoid any bruises on the fruit skin.[20] The scientific work of analyzing fruit skins demanded newly compliant human bodies, a new ecology of fruits and laborers; this was a landscape of reformed democracy, new markets, and racial fault lines.

A PEOPLE'S MONOPOLY

The California Fruit Growers Exchange—the citrus cooperative founded in 1905 but whose origins could be traced to local initiatives since the 1880s—had, as first building blocks, the local associations for the joint funding of packinghouses. These were expensive facilities that, with their sophisticated grading machinery, prepared fruit for shipment, putting in contact western growers with markets in the main cities of the East Coast.[21] Each local association owned a packinghouse in which the fruit was washed, brushed, dried, graded, packed, and labeled. By pooling fruit from different growers, packinghouses increased the bargaining power of the growers in relation to shippers and eliminated middlemen, a central issue for producers three thousand miles away from their urban markets. The co-op demonstrated that the twelve thousand to fifteen thousand California citrus growers, the large majority owning small groves with an area between five and ten acres, when properly organized in associations, could actually prosper in a market economy and face the power of railway monopolies—the octopus that embodied the evils of US capitalism at the turn of the century.

It is thus important to understand that Powell's advice to have gangs of laborers controlled by packinghouse management meant no less than putting in the hands of the growers' co-op the task of recruiting workforces to labor in the orange groves. According to the statistics of the growers organizations, in 1912, there were twenty-five thousand laborers employed in the citrus orchards. Instead of different individual growers establishing contracts with laborers, this task was now to be centralized by the co-op if one wished to avoid the decay of citrus and its associated losses. According to Powell, only the expansion of the capacity of the co-op to intervene in every stage of the industry would sustain the high levels of income experienced by California orange growers since the last decades of the nineteenth century.

Shortly after its formation in 1905, the California Fruit Growers Exchange was harvesting, packing, shipping, marketing, and financing 70 percent of the region's orange crop. It consisted of a three-tiered nonprofit organization formed by 201 local packing associations; twenty-five district exchanges working as selling agencies for the packing associations; a central exchange that coordinated the rest of the structure, with sales agents distributed throughout the country; as well as field agents responsible for the inspection of the quality of the fruit at key sites of distribution.[22] The agents sent daily telegrams on the status of the market to the office of the central exchange located in downtown Los Angeles, which in turn produced a daily report of the citrus situation nationwide. These reports were distributed among the local associations and district exchanges, determining the destination of citrus shipments.

This bureaucratic structure of the citrus growers' cooperative had obvious resemblances with those of large corporations at the time, such as Westinghouse or the Southern Pacific Railroad, leading many authors to describe it simply as the arrival of big business into agriculture. But as Charles Postel has recently demonstrated in his important revisionist history of the Populist Movement and its struggle against the monopolies that dominated US capitalism of the Gilded Age, what was at stake in the formation of farmers' cooperatives was the potential "to fight capital with capital."[23] In Texas, when facing accusations that cotton farmers were forming a new monopoly, populist leaders replied, "If it is a monopoly we shall create a grand one. It will be a philanthropic monopoly. It will distribute wealth among the people."[24] The promoters of the co-ops were thus self-conscious about the apparent confusion of institutional forms. This was especially true for citrus: "Even in California many who have not had occasion to acquaint themselves with its real nature and purpose regard the exchange as the orange trust instead of what it is, a democratic group of cooperators."[25]

Historians haven't made much of a case of such claims insisting on the simple fact that large growers always dominated the California Fruit Growers Exchange, which allegedly indicates that democratic talk was never more than empty rhetoric hiding cruder capitalist intentions.[26] But the point is that citrus growers went a long way to explain how their cooperative formation was an alternative to existent capitalist practices while acknowledging differences in the community of growers. In each local packing association, voting power and contributions to the acquisition of common equipment were determined as a function of the number of acres each grower possessed. It was understood that the blind application of the principle of "one man, one vote" couldn't promote an equitable community meant to bring together fairly owners of one hundred acres and of five acres. The large sums of capital involved in acquiring packing machinery or a precooling plant determined that larger contributors would only be part of the undertaking if their higher financial stakes were somehow recognized. But while important matters for the life of the association, such as the election of its manager, were thus decided on the base of differentiated voting power determined by acreage, most daily decisions were taken on simple counting of individual *ayes* and *nays*. As stated in one self-interested account of the functioning of the exchange, "If any member feels that his influence or power in the association is so insignificant that he is in no way able to affect its general policy he is quite likely to sever his connection with it." Or, more bluntly, the cooperative "shall be under democratic control.... No individual or small group of individuals because of its preponderating importance in the organization should cause it to become a miniature despotism or oligarchy."[27]

The challenges of balancing equity with equality in a modern social organization where technical expertise was crucial were made patent in the annual meeting of the association organized every September. The meeting was a big local social event, alternating reading of reports and business discussions with collective dinners put together by growers' wives and with the cheerful presence of their children. There is no doubt that the managers, always elected by larger growers, were the protagonists: they were the ones who, from early morning, set the agenda with their annual reports, informing the collective assembly of tonnage of fruit shipped, cost of handling, the average net price of a box of oranges, or losses caused by yet another pest afflicting the groves. But in the afternoon sessions, business could be introduced directly from the floor, with any grower, independent of size, able to express grievances and concerns on how the collectivity was being operated: "The managers of the association and of the district exchange often come in for a thorough grilling, being called to task for past shortcomings and exhorted as to future conduct. Policies recommended by the management are dissected and scrutinized. Counter policies are proposed by members."[28]

The commercial success of the California citrus co-op probably contributed to the historical forgetting of the social experimentation involved in its formation. It is not common among historians of capitalism to identify capitalist success with democratic ventures. And in fact the citrus industry of Southern California grew at a rate unparalleled by any agricultural product in the world. Fueled by the cheap credit provided by the financial institutions of San Francisco and Sacramento that came out of the Gold Rush, the restrictive tariffs imposed on imports from Southern Europe, as well as by increases in productivity that almost doubled tree yielding, shipments of California citrus increased 580 percent between 1894 and 1914.[29] In Riverside, the center of the industry, and as early as 1895, residents boasted the highest per-capita income in the United States. The proliferation of banks, theaters, mansions, tree-lined streets, churches, and colleges in the communities of the citrus belt around Los Angeles—from Pasadena to Ontario, Riverside to Redlands—praised throughout the landscape the growers' accomplishments. Instead of Lorenzetti's fresco, the image of the virtuous community was now mass distributed on orange crates as well as through the lavishly illustrated pages of the *Citrograph*, the mouthpiece of the citrus industry, in which pictures of growers' bungalows and public parks were published side by side with scientific advice by the likes of G. Harold Powell on citrus pathologies or fertilization techniques.[30] While multiacre bonanza farms and their highly mechanized operations were driving small farmers out of business in the Great Plains, Southern California's modern specialty crop agriculture actualized the Jeffersonian democratic rural ideal through a combination of co-ops, state intervention, and scientific research.

THE SURVEY AS COMMUNITY-BUILDING TOOL

Powell was the perfect embodiment of such a combination. His scientific survey of the citrus country would prove to be an exemplary case of what cultural anthropologists would recognize as participant observation, transforming him into a crucial figure of the local cooperative culture. His investigations demanded a close collaboration with the growers, who gave him full access to their groves and who provided him with laboratory facilities at their ranches.[31] Powell was offered an even more intimate knowledge of his research objects through his continuous presence at the growers' dining table at the Mission Inn in Riverside.[32] The luxurious hotel was the center of citrus-elite sociability and would attract in the following decades Hollywood stars, oil barons, and US presidents. This authentic theme park of Mission-style revival with its *patio de las fuentes* (court of the fountains), balconies enclosed in old Spanish railings, and Spanish art gallery and collection of church bells, a setting that consolidated many expressions of the reverie of the new citrus landscape, was Powell's home during his three years of fieldwork.[33] It didn't take long for Powell to succumb to the seduction of Southern California. By 1910, he was moving with his family to Pasadena to assume the position of no less than general manager of the California Fruit Growers Exchange.

In the following years, this Southern California citrus cooperative would be praised as exemplary of the reforms advanced by the Progressive Movement. Namely, it fulfilled the principles of the Country Life Commission launched by president Theodore Roosevelt in 1908 to bring progressivism to US farmlands as stated in its guiding motto: "Better business methods, better farming, and better living." It is telling that no other author than Powell would be chosen to write a volume on the report "Cooperation in Agriculture" for the Rural Sciences Series edited by Liberty Hyde Bailey, the prophet of the Country Life Movement.[34] In good Progressive jargon, Powell used the citrus co-op of which he was now the general manager to inform readers about how the association raised the economic efficiency of an entire community. Protection from wholesale dealers, better handling of the fruits, fumigation procedures, protection of the orchards against frost, and pruning were all mobilized to make the case that farmers could only thrive in modern times by embarking as part of a scientifically managed co-op.

The importance of oranges in this scientific version of back-to-the-land ideology, publicized by a movement led by urban elites invested in uplifting Country Life, shouldn't surprise us, considering that Powell had started his career in horticulture as a distinguished disciple of Bailey at the Cornell College of Agriculture. It was at Cornell that Bailey transformed horticulture into a respectable scientific discipline, convincing the New York State Board to build the first campus horticultural building in the country.[35] In Bailey's view, scientific horticulture and the reform of Country Life were one and the same. School and community gardens were to become central elements of sociability in reinvigorated rural communities, informed by scientific practices disseminated by nature study of which he was one of the main promoters. In fact, Bailey's slim volume *The Nature-Study Idea* (1903) would contribute to making him one of the better-known figures in the life sciences in the country.[36] The above mentioned contemporary efforts by John Dewey at the Chicago Laboratory School, bringing together education, research, and community building through the practice of horticulture in the school's garden, drew direct inspiration from Bailey's nature study.[37] As proclaimed by the citrus landscape of Southern California, horticultural experimenting meant experimenting with democracy.

And school gardens were not the only tools that Dewey shared with Bailey to reinvigorate US democracy. In his *Reconstruction in Philosophy* (1920), Dewey advanced the notion that "organization is never an end in itself."[38] He took for granted that "individuality is not originally given but is created under the influences of associated life," only to make the more important point that philosophers should address the question of what sort of individuals were created by specific forms of social organization: "Just what response does this social arrangement ... evoke? ... Does it release capacity? If so, how widely? Among a few, with a corresponding depression in others, or in an extensive and equitable way?"[39] Dewey thus advocated for "inquiries about every institution of the community, ... specific inquiries into a multitude of specific structures and interactions."[40] These social-philosophy enquiries blurred the lines between philosophy and engineering by producing a criticism of society that, in typical Progressive way, followed "the methods of science."[41] In fact, they seem to share the same nature of the surveys that Bailey promoted from his post at Cornell and that would constitute the model for the report on the US countryside at the national scale produced by the Country Life Commission.[42]

Bailey was emphatic about the democratic dimensions of agriculture surveys. Their aim was no less than "to tie the community together.... Apple-growing would not be distinct from wheat-growing, or church work from schoolwork, or soil types from the creamery business, or politics from home life. The vicinage would be presented to the citizen as a whole. We are to build the life of every community on the fact of that community."[43] Bailey's reference to apple growing was not casual, since the first of such surveys was undertaken at Cornell in 1890, with a report on the condition of fruit growing in western New York. The horticulture survey started as a study on the causes of rural economic depression and the ways of overcoming it, soon to evolve into an encompassing portrait of an entire community. The survey was portrayed by

Bailey as the guarantee that collective action always arose from knowledge of the facts of the community and not from unreliable opinions; in other words, that decisions about the life of the community were to be taken based on science and not on ideology.

In particular, Bailey's visions informed the 1914 federal Smith-Lever Act instituting extension services as part of the mission of the land-grant colleges and aiming at improving rural life through the systematic dissemination of scientific knowledge. Coming back to our Southern California citrus story, they were also the obvious direct source of inspiration for Powell's "The Decay of Oranges While in Transit from California," the survey that led to the "rationalization" of citrus handling and to the consequent election of the USDA scientist as general manager of the citrus co-op.

ROOTSTOCKS

Among the long list of Bailey's disciples at Cornell was also Herbert John Webber, whose name is a common presence in the historiography of genetics as one of the key figures in introducing Mendel's heredity principles in the United States.[44] He had arrived in Ithaca in 1907 to direct its plant-breeding department, having become the director of the college of agriculture in 1910. He didn't last long in the post, for two years later he was leaving what was then one of the most prominent positions in the life sciences in the United States to accept UC's invitation to lead its Citrus Experiment Station at Riverside.[45]

The citrus growers had been lobbying the UC regents for many years for a scientific institution that would answer directly to their anxieties about unstable commercial and orange-growing conditions.[46] The College of Agriculture at Berkeley repeatedly denied the proposal for organizing an entire institution around a single commodity, but Powell's work on blue mold seemed to have changed the regents' opinion and, in 1906, a citrus experiment station was inaugurated in a small plot of land on the eastern slope of Mount Rubidoux in Riverside. In 1918, already under the leadership of Webber, the station would move to its flamboyant new building, adapting the mission style to laboratory architecture, overseeing some 480 acres where the UC Riverside campus is located today. Although not as centrally located as before, it was clear that the station had the growers as its main constituency. Webber, following the example of Powell, and of their mentor Bailey, would constantly remind listeners of the role of research in strengthening the cooperative movement.[47] The difference now was that he could count on the collaboration of fourteen other full time researchers, six laboratory assistants, fourteen foremen, and an unidentified number of farm laborers. A battery of disciplines scrutinized citruses with research distributed across departments of plant pathology, physiology, agricultural chemistry, entomology, plant breeding, soils, and orchard management.

Webber was particularly interested in rootstock investigations, undertaking painstaking experiments that ran for more than twenty years, highlighting reproduction mechanisms, disease resistance, and graft compatibilities with various commercial citrus varieties.[48] Rootstocks assumed a crucial role in California orchards that produced mainly two citrus varieties: Washington Navels and Valencias. These susceptible varieties were budded on rootstocks from resistant varieties, mostly sour orange, protecting the citrus tree from menacing diseases such as gummosis. Although the UC Citrus Experiment Station breeders had an interest in introducing new commercially promising varieties into California orchards, it should be stressed that their first concern was to keep the properties of trees producing Navels and Valencias. More than research directed at innovation, this was a matter of maintenance. To guarantee the uniform high yields on which the co-op members relied, it was necessary that the propagation procedures of rootstocks guaranteed uniform specimens that reproduced true to type. This was the core issue of Webber's long-term rootstock experiments.

Citrus' seedlings originate both from sexual and nonsexual reproduction. Webber's research demonstrated that the best rootstocks were those in which a large proportion of the seedlings came from embryos produced through nonsexual reproduction by the budding of the cells surrounding the egg cell in the flower of the mother parent (nucellar embryony). Contrary to the seedlings resulting from fecundation, these specimens showed much higher uniformity, since they only originated from a single parent. Only varieties producing 75 percent to 100 percent of seedlings in this way were considered for their use as rootstocks. As Webber stated, "The non-sexually produced seedlings come very true to type of the seed parent and are much more uniform in constitution and type than sexually produced seedlings. This phenomenon is therefore an important one in the production of citrus seedlings for budding."[49]

Now, his research also showed that variations still occur among those nucellar seedlings, namely due to mutations. A considerable number of small and off-type individuals, when budded to standard fruit varieties, produce unsatisfactory dwarfed orchard trees and should be discarded. Webber thus advocated for a method of nursery selection leading to the production of uniform high-yielding orchards. "It consisted first in the elimination of small and off-type seedlings in the seed bed, and again in the nursery just before budding; and second in the discarding of small budded nursery trees at the time of transplanting to the orchard."[50] The many local nurseries that supplied the Southern California citrus growers were to follow such a procedure, selling only trees whose rootstocks had been standardized.

Webber himself had first coined the modern notion of clone in an article published in *Science* in 1903.[51] Making use of the Greek word *clon*, meaning a twig, spray, or slip, such as is broken off for propagation, he defined clones as groups of plants propagated by the use of any form of vegetative parts such as bulbs, tubers, cuttings grafts, buds, etc., and which are simply parts of the same individual seedling. An interesting point for our discussion is how Webber mentioned that the plants grown from such vegetative parts are not individuals in the ordinary sense but are simply transplanted parts of the same individual, and in heredity and in all biological and physiological senses such plants are the same individual. Or, in other words, in its efforts to standardize orange orchards, the citrus co-op was built around a few cloned organisms, binding together its several thousand human members.

BUDDING AND BUDDIES

Let us insist on the importance of cloning practices for the maintenance of the Southern California social order by turning now to the work of Archibald D. Shamel on budding variations. A physiologist from the USDA Bureau of Plant Industry, Shamel arrived in Riverside in 1909, also at the request of the growers challenged by the growing variations found in their orchards. Although Washington Navels and Valencias were reproduced through budding, the number of undesirable trees was clearly increasing, producing "irregular, light crops of inferior quality."[52] Shamel thus started research that would occupy him for the next twenty-seven years, studying the source of strains originating as bud mutations and unintentionally propagated by nurserymen and growers, as well as implementing a scheme to eliminate so-called drone trees.[53]

Shamel undertook typical physiology studies of flower and fruit characteristics, showing, for example, that several individual trees had no less than nine strains of the navel orange on different limbs, all of them arising as bud sports in these trees. By 1919, Shamel's survey of Southern California orchards had identified a total of thirteen strains of the Washington Navel, twelve strains of the Valencia orange, six strains of the Marsh grapefruit, eight strains of the Eureka lemon, and five strains of the Lisbon lemon varieties. The consequences for the fortune of the growers was obvious, as asserted by the results concerning the Washington Navel variety:

"About 25% of the total number of trees studied in the original orchards in which these investigations have been conducted were found to be of undesirable strains having consistently low yields, or bearing fruits of poor quality, or both, such as those of the Australian, Unproductive, Corrugated, Pear-Shape, Sheep-nose, Flattened, Dry, and other inferior strains."[54] Shamel's systematic survey of the orchards traced back each strain to bud variations accidentally propagated due to a lack of knowledge of their significance "in the work of maintaining the citrus varieties."

The survey translated the Southern California citrus landscape into a genealogical map of different strains, identifying important variations where the lay eye saw only uniform rows of orange trees. Of course, the work demanded easy access from growers to their orchards in order to observe individual trees, take photographs, and cut fruits for further examination in the laboratory of the citrus experiment station. But the degree of collaboration was in fact much more intense, leading to the transformation of the orchard into a controlled, homogeneous space able to produce scientific facts. At the center of such process was the "performance record" in which the grower registered the number and quality of fruits produced by each of his individual trees. The best strains of each variety were expected to also be the heaviest producers of fruit. The aim of the records was to locate the drone trees present in the orchard; identify the superior trees from which bud wood could be obtained; recognize trees in need of care (cutting out limb spots, disease treatments, etc.); and compare the effect of cultural treatments or any other experimental tests. In other words, Shamel's experiment transformed the orchard into a lab, eliminating lab/field division, with every grower participating in the experiment. By the end of the 1910s, around fifty thousand acres of California citrus orchards kept performance records.[55]

The first building block of this bureaucracy effort was to give a number to each tree, indicating its number of the block of the orchard, the number of the row in the block, and the position of the tree in the row. Every tree had now painted on its trunk large, easily legible figures for its quick identification. When picking, boxes were to be distributed to the individual trees instead of the old method of using box rows, and all the fruit from each tree was to be placed in boxes at its base. The performance record thus also carried important consequences for the pickers' practices: "Each picker's work is always open to inspection. With one picker on a row the natural tendency is to induce the slower pickers to keep up with the faster workers. Inasmuch as the field boxes are near the tree being picked, this arrangement does away with the necessity for each man walking with his filled picking sack from the tree to the box row, as was formerly the case, and in this way saves considerable time."[56] Performance records were not only good to eliminate drone trees and keep the homogeneity of the orchard. They also served to do away with slow pickers and guarantee streamlined labor practices.

The foreman of each picking crew now had the added task of filling up the performance record forms. Each day, before the fruit was transported to the packinghouse, the foreman recorded in a field notebook the number of boxes picked from each tree, weighing each partly filled box. The form also included the name of the variety, the date of picking, and the tree number, as well as notes on the apparent quality of the fruits (first grade, second grade, culls) and of any unusual tree condition.

Considering the importance of uniformity of production for the good working of a co-op pooling the fruit of several thousand growers, each being paid as a function of the volume of citrus delivered to the packinghouse, the attention paid by the growers to Shamel's work, who named his large-scale experiment "Cooperative improvement of citrus varieties," is not surprising. In 1917, the co-op established a department of bud selection, aimed at securing bud wood from superior performance-record trees and distributing it to propagators.[57] The department having access to the performance records gathered by individual growers identified the best trees in each orchard from which buds were to be taken: "The selection of parent trees has been guided by an intimate knowledge of the trees of the variety gained through systematic individual-tree record work." The bud sticks from each parent tree were kept in separate bundles. A tag with a serial number was attached to each bundle and a duplicate tag with the same serial number, the number of the tree from which the buds were cut, and the name of the propagator to whom the buds were to be sent was filed in the co-op bud-selection department. This information, together with the individual-tree records, made possible the tracing of any progeny in a nursery to the parent tree in the orchard, increasing transparency for growers purchasing their trees from nurserymen. As the co-op was responsible for securing and distributing the buds, these were supplied to propagators at cost determined by the "payment for the buds to the owners of the parent trees, the assembling, tabulating, and studying of extensive individual-tree data, the selection of the superior parent trees, collecting information as to the behavior of the buds and the trees grown from them, and the survey of new orchard areas for the location of additional parent trees." From 1917 to 1935, the co-op had distributed no less than 1,402,950 selected buds from superior strains of the Washington Navel orange and 2,338,004 of the Valencia.[58] If, on the one hand, Shamel had access to a large number of records for his studies of variation in citrus, thus basing his work on the co-op structure, on the other hand, the co-op expanded its realm of intervention by guaranteeing that each of its members was cultivating uniform, standardized trees—the clones of the best performers. Cloning practices were essential for sustaining and reproducing the growers' co-op.

CONCLUSION

A. D. Shamel's easy access to and deep knowledge of every orchard in Southern California placed him in a privileged position to become one of the pioneers in documenting the changes occurring in the landscape.[59] His camera was directed not only at individual trees in the orchard, but also at the gardens surrounding the bungalows of citrus growers and, maybe more unexpectedly, at farmworkers' housing. From 1918 onward, he famously produced a long series of photographic essays published during a decade in the California *Citrograph* on housing projects for workers at the Limoneira Company, Rancho Sespe, and the Upland Citrus Association.[60] In such essays, Shamel shows a particular interest in site selection, construction standards, dispositions of interior space, porches, gardens, and even flower boxes. The intention was clear: promote in the citrus industry a stable workforce, mainly of Mexican origin, in contrast to the migrant labor that characterized most agriculture operations in the rest of the state. His photographs of citrus villages taken in the early morning suggested the harmony between architecture details and the natural setting, reiterating the virtuous nature of the citrusscape. Adding to the lavish images of the Mission Inn, the tree-lined Magnolia avenue at Riverside, and the bungalows of successful growers, Mexican farmworkers were also integrated into the scene, occupying cottages and dormitories equipped with running water, gas, and electricity; the setting was also shaded by eucalyptus, sycamore, and oak trees. Shamel's concern with a small space available for gardening, even in the form of modest flower boxes, reiterates the idea of the moral effects of horticulture in building a stable community.

Although working conditions in the citrus orchards around Riverside were better than in most other branches of agriculture, thus dismissing the too-homogeneous descriptions associated with farmwork in California, it doesn't take much scholarly sophistication to debunk Shamel's idyllic paternalistic views.[61] There is no dearth of good literature on the subject highlighting the racial fault lines of the modern citrus version of the Garden of Eden.[62] While not ignoring the intense communal life of Mexican colonias and the role of farmworkers in defining the urban culture of the entire

Los Angeles area, the separation between communities instituted a US equivalent of the apartheid regime. The beautiful villages portrayed by Shamel and constantly reproduced in the media by California boosters were indeed fully segregated communities in which racial lines were constantly monitored. If neighborhoods, schools, theaters, and sporting events reproduced racial divisions on a daily basis, the crucial separation was instituted by the impossibility of brown and yellow people integrating the all-white egalitarian citrus co-op. Before Mexicans came to dominate the workforce of the region, here is how local press described Japanese workers:

> His color sets him off from the rest of us so far as to make of him a marked man. It may be urged that this ought not to make any difference, that a man is a man, no matter what the tint of his skin. Granted—but this is a democracy, and people must be taken as they are. We cannot fraternize with colored peoples as we do with each other. ... We cannot do the business of democracy with people so strongly set off from us in racial character. Their presence among us in great numbers raises the most explosive questions—questions of sex, marriage, school life, church life, business life, traveling problems, questions of all sorts of mingling.[63]

In other words, the democratic experience of the citrus co-op was only possible, following its promoters, through the drawing of racial fault lines. It is not that people of color were not admitted into an already existent co-op, but that the latter was built on racial distinctions between white subjects as scientifically informed citizens experimenting with US democracy and workers of color formulated as objects of scientific optimization through the researches of the likes of Powell and Shamel. The very same insistence on the role of Americanization classes throughout the citrus belt towns in order to form a second generation of loyal, grateful, English-speaking workers suggests the continuity between Shamel's scientific undertakings and his social concerns. What has normally been perceived as two distinct activities should actually be seen as one single concern with the maintenance of the sociability built around citrus. Standardizing farmworkers' housing or cloning citruses formed a continuum of reproducibility practices.

If, in the Progressive Era, the citrus co-op was celebrated as an exemplary growers' monopoly countering the monopoly of railways, in the New Deal years it was commonly denounced for such undemocratic manners among growers' by wageworkers demanding better conditions. When in 1936 a strike of citrus workers was violently repressed in Orange County by local deputies and vigilantes, Carey McWilliams aptly changed the name of the co-op from Sunkist into Gunkist.[64] There can be no doubt about the noir dimension of Shamel's beautiful photos. The comparison with similar violent practices repressing farmworkers' organizations by the paramilitary black shirts of Mussolini may suggest an uncomfortable revelation of a kernel of fascism within the Californian democratic dream.

Such a line of thought was the one pursued by Theodor Adorno and Max Horkheimer in *Dialectic of Enlightenment*, written, not coincidentally, during their Californian exile from Hitler's regime.[65] If they made no direct reference to the citrus orchards around them, they were nevertheless quick to denounce the standardized bungalows that sprouted in Los Angeles. According to the scandalous perception of the founders of Critical Theory, the dwellers of the Southern Californian metropolis were "virtually already Nazis." Sustaining that "technical rationality today is the rationality of domination," Adorno and Horkheimer took the general process of technical standardization as making democratic and authoritarian regimes indistinguishable from one another through the "infection [of] everything with sameness." This powerful critique would appeal to many scholars in the second half of the twentieth century interested in scrutinizing the nature of modernity. Their task was now clear: to identify the pervasive presence of fascism in democratic societies.

Were we to follow Adorno and Horkheimer too closely, the citrus story would constitute no more than a case study illustrating the more general phenomenon of standardization producing social domination through reproduction of sameness.[66] Such critical undertaking would blind us to the actual historical dynamics leading to the standardization of citrus. As we saw, the Washington navels and Valencias grown in California came into being as embodying an alternative to the unregulated capitalism of the Gilded Age, sustaining an enlarged community of growers. Science and technology didn't make everything the same: while wheat bonanza farms in the Midwest grew larger and larger with increased mechanization, in Southern California, a cooperative community of thousands of white horticulturalists thrived on small ten-acre orchards grown on new cloning practices. Closer to the vision of John Dewey mentioned above, one may conclude by looking at the citrus scientific co-op that standardization promoted "the movement toward multiplying all kinds and varieties of associations," and in so doing, enabling the social experimentation characteristic of a democratic society.[67] What the progressive vision of Dewey didn't highlight was that the experiment "of multiplying effective points of contact between persons" also institutionalized the separation between growers and workers, whites and nonwhites, making race constitutive of US democracy.

NOTES

1. Bruno Latour, "From Realpolitik to Dingpolitik," in *Making Things Public: Atmospheres of Democracy*, eds. Bruno Latour and Peter Weibel (Cambridge, MA: MIT Press, 2005), 14–44. For other interpretations of the fresco, see Quentin Skinner, "Ambrogio Lorenzetti: The Artist as Political Philosopher," *Proceedings of the British Academy* 72 (1986): 1–56; "Ambrogio Lorenzetti's Buon Governo Frescoes: Two Old Questions, Two New Answers," *Journal of the Warburg and Courtauld Institutes* 62 (1999): 1–28.
2. For examples of such hybrid fora, see Michel Callon, Pierre Lascoumes, and Yannick Barthe, *Agir dans un monde incertain: essai sur la démocratie technique* (Paris: Editions du Seuil, 2001).

3. For the relation between pragmatism and STS, see Noortje Marres, "The Issues Deserve More Credit: Pragmatist Contributions to the Study of Public Involvement in Controversy," *Social Studies of Science* 37 (2007): 759–80
4. John Dewey, *The Public and Its Problems* (Carbondale: Southern Illinois University Press, [1927] 1981), 365.
5. See, namely, John Dewey, *Reconstruction in Philosophy* (Boston, MA: Beacon Press, [1927] 1957).
6. For a more detailed discussion of the problems of using philosophical concepts of STS without historicizing them, specifically Latour's use of Heidegger's "Thing," see Tiago Saraiva, *Fascist Pigs: Technoscientific Organisms and the History of Fascism* (Cambridge, MA: MIT Press, 2016): 237–42.
7. Sally Gregory Kohlstedt, "Nature, Not Books: Scientists and the Origins of the Nature-Study Movement in the 1890s," *Isis* 96 (2005): 324–52; Ben A. Minteer, *Landscape of Reform: Civic Pragmatism and Environmental Thought in America* (Cambridge, MA: MIT Press, 2006), 17–50.
8. Katherine Camp Mayhew and Anna Camp Edwards, *The Dewey School: The Laboratory School of the University of Chicago, 1896–1903* (New York: D. Appleton-Century, 1936).
9. On the notion of experimental systems in the history of science, see Hans-Jörg Rheinberger, *Towards a History of Epistemic Things: Synthesizing Proteins in Test Tube* (Stanford, CA: Stanford University Press, 1997).
10. There is very abundant literature on this historical context. Particularly valuable are Douglas Cazaux Sackman, *Orange Empire: California and the Fruits of Eden* (Berkeley: University of California Press, 2005); Matt Garcia, *A World of Its Own: Race, Labor, and Citrus in the Making of Greater Los Angeles, 1900–1970* (Chapel Hill: University of North Carolina Press, 2001); Gilbert G. Gonzalez, *Labour and Community: Mexican Citrus Worker Villages in a Southern California County, 1900–1950* (Urbana and Chicago: University of Illinois Press, 1994); Margo McBane, *The House That Lemons Built: Race, Ethnicity, Gender, Citizenship, and the Creation of a Citrus Empire, 1893–1919* (PhD diss., University of California, Los Angeles, 2001); George L. Henderson, *California and the Fictions of Capital* (New York: Oxford University Press, 1999); Ronald Tobey and Charles Wetherell, "The Citrus Industry and the Revolution of Corporate Capitalism in Southern California, 1877–1944," *California History* 74, no. 1 (1995): 6–21; David Vaught, "Factories in the Field Revisited," *Pacific Historical Review* 66, no. 2 (1997): 149–84; H. Vincent Moses, "G. Harold Powell and the Corporate Consolidation of the Modern Citrus Enterprise, 1904–1922," *Business History Review* 69 (1955): 119–55; and Alan L. Olmstead and W. Paul Rhode, "The Evolution of California Agriculture 1850–2000," in *California*

Agriculture: Dimensions and Issues, ed. Jerome B. Siebert (Berkeley: University of California, 2004): 1–28.

11. Andrew Jewett, *Science, Democracy, and the American University: From the Civil War to the Cold War* (New York: Cambridge University Press, 2012); David F. Noble, *America by Design: Science, Technology, and the Rise of American Capitalism* (New York: Alfred A. Knopf, 1977).

12. Scott J. Peters and Paul A. Morgan, "The Country Life Commission: Reconsidering a Milestone in American Agricultural History," *Agricultural History* 78 (2004): 289–316; Clayton S. Ellsworth, "Theodore Roosevelt's Country Life Commission," *Agricultural History* 34 (1960): 155–72.

13. Such interpretation is in accordance with institutionalist scholars' important proposals, urging us to put aside simplistic oppositions between the federal state and civic associations, and understand instead the relation between the two as one of interdependence. Theda Skocpol, Marshall Ganz, and Ziad Munson, "A Nation of Organizers: The Institutional Origins of Civic Voluntarism in the United States," *American Political Science Review* 94, no. 3 (2000): 527–46.

14. Garcia, *A World of Its Own*; Gonzalez, *Labor and Community*; McBane, *The House*; Tomas Almaguer, *Racial Fault Lines: The Historical Origins of White Supremacy in California* (Berkeley: University of California Press, 1994).

15. Morton Keller, *Regulating a New Society: Public Policy and Social Change in America, 1900–1933* (Cambridge, MA: Harvard University Press, 1994); William A. Link, *The Paradox of Southern Progressivism* (Chapel Hill: University of North Carolina Press, 1992); Elizabeth Grace Hale, *Making Whiteness: The Culture of Segregation in the South 1890–1940* (New York: Pantheon, 1998); George M. Fredrickson, *White Supremacy: A Comparative Study in American and South African History* (Cambridge: Cambridge University Press, 1982); Steven Hahn, *A Nation without Borders: The United States and Its World in an Age of Civil Wars, 1830–1910* (New York: Penguin, 2016), 476–85; Almaguer, *Racial Fault Lines*.

16. H. Vincent Moses, "'The Orange-Grower Is Not a Farmer': G. Harold Powell, Riverside Orchardists, and the Coming of Industrial Agriculture, 1893–1930," *California History* 74 (1995): 22–37; Moses, "G. Harold Powell."

17. The following paragraphs are based on Powell's report: G. Harold Powell, "The Decay of Oranges while in Transit from California" (Washington, DC: USDA, 1908).

18. Powell, "The Decay," 27.

19. Powell, 23.

20. Powell, 27–29.

21. W. W. Cumberland, *Cooperative Marketing: Its Advantages as Exemplified in the Californian Fruit Growers Exchange* (Princeton, NJ: Princeton University Press, 1917), 47; Irwin W. Rust and Kelsey B. Gardner, *Sunkist Growers Inc.: A California Adventure in Agriculture Cooperation* (Washington, DC: USDA, 1960); Henderson, *Fictions of Capitalism*, 64–70.
22. Henderson, *Fictions of Capitalism*, 64–65.
23. Charles Postel, *The Populist Vision* (New York: Oxford University Press, 2007).
24. Postel, *Populist Vision*, 103.
25. Cumberland, *Cooperative Marketing*, 59.
26. Henderson, *Fictions of Capital*; Moses, "The Orange-Grower."
27. Cumberland, *Cooperative Marketing*, 13.
28. Cumberland, 76.
29. Henderson, *Fictions of Capital*; Rust and Gardner, *Sunkist Growers Inc.*
30. Sackman, *Orange Empire*.
31. Powell, "Decay of Oranges."
32. Lawrence Clark Powell, *Portrait of my Father* (Santa Barbara, CA: Capra Press, 1986).
33. On Southern California landscape and Spanish colonial revival, see Ian Tyrrell, *True Gardens of the Gods: Californian-Australian Environmental Reform, 1860–1930* (Berkeley: University of California Press, 1999).
34. G. Harold Powell, *Cooperation in Agriculture* (New York: Macmillan, 1918).
35. Minteer, *Landscape of Reform*, 17–50.
36. Kohlstedt, "Nature, Not Books."
37. Minteer, *Landscape of Reform*.
38. Dewey, *Reconstruction in Philosophy*, 206.
39. Dewey, 197–98.
40. Dewey, 197–98.
41. Richard Rorty, "Overcoming the Tradition: Heidegger and Dewey," *The Review of Metaphysics* 30, no. 2 (1976): 280–305, specifically 291.
42. Liberty Hyde Bailey, *The Survey-Idea in Country Life Work* (New York: MacMillan, 1911).
43. Bailey, *Survey-Idea*, 14–15.
44. Diane B. Paul and Barbara A. Kimmelman, "Mendel in America: Theory and Practice, 1900–1919," in *American Development of Biology*, ed. Rainger et al. (Philadelphia: University of Pennsylvania Press, 1988), 281–310.
45. Herbert J. Webber, "Autobiographical Sketches," Herbert J. Webber papers, Box 1.2, Collection 059, Riverside Libraries, Special Collections & Archives, University of California, Riverside.
46. Harry W. Lawton and Lewis G. Weathers, "The Origins of Citrus Research in

California," in *The Citrus Industry*, vol. 5, ed. Walter Reuther et al.(Berkeley: University of California Press, 1989), 5, 281–335.

47. Herbert J. Webber, *What Research Has Done for Subtropical Agriculture: Achievements of the Citrus Experiment Station* (Berkeley: University of California, 1934).
48. Herbert J. Webber, "Variations in Citrus Seedlings and Their Relation to Rootstock Selection," *Hilgardia* 7 (1932): 1–79; Herbert J. Webber, "The Improvement of Root-Stocks Used in Fruit Propagation," *Journal of Heredity* 11 (1920): 291–99; Herbert J. Webber and J. T. Barrett, "Root-Stocks Influence in Citrus," Ninth International Horticultural Congress, London, 1930.
49. Webber, *What Research Has Done*, 16.
50. Webber, 16.
51. Herbert J. Webber, "New Horticultural and Agricultural Terms," *Science* 18 (1903): 501–03.
52. A. D. Shamel, "Cooperative Improvement of Citrus Varieties," in *Yearbook of the Department of Agriculture* (Washington, DC: USDA, 1919), 249–75, specifically 250.
53. A. D. Shamel, "Citrus Fruit Improvement: A Study of Bud Variation in the Washington Navel Orange" (Washington, DC: USDA, 1918); A. D. Shamel, "Citrus Fruit Improvement: A Study of Bud Variation in the Valencia Orange" (Washington, DC: USDA, 1918).
54. Shamel, "Cooperative Improvement," 255.
55. Shamel, 255.
56. Shamel, 262–64.
57. Shamel, 265–75.
58. Hamilton P. Traub and T. Ralph Robinson, "Improvement of Subtropical Fruit Crops" (Washington, DC: USDA, 1937), 784–85.
59. Anthea M. Artig, "In a World He Has Created: Class Collectivity and the Growers' landscape of the Southern Californian Citrus Industry, 1890–1940," *California History* 74 (1995): 100–11.
60. Richard Steven Street, *Everyone Had Cameras: Photography and Farmworkers in California, 1850–2000* (Minneapolis: University of Minnesota Press, 2008); A. D. Shamel, "The Esthetic Side of Orange Growing in the South West," *Citrograph* (November 1929).
61. Olmstead and Rhode, "Evolution of California," 1–28.
62. See, namely, Garcia, *A World of Its Own*; Gilbert. G. Gonzalez, "Labor and Community: The Camps of Mexican Citrus Pickers in Southern California," *The Western Historical Quarterly* 22/3 (1991): 289–312.
63. Quoted in Margo McBane, "Whitening a Californian Citrus Company Town:

Racial Segregation Practices at the Limoneira Company and Santa Paula, 1893–1919," *Race/Ethnicity* 4, no. 2 (2011): 211–33, specifically 213.

64. Carey McWilliams, "Gunkist Oranges," *Pacific Weekly*, July 20, 1936.
65. Max Horkeimer and Theodor W. Adorno, *Dialectic of Enlightenment: Philosophical Fragments*, ed. Gunzelin Schmidt Noerr, trans. Edmund Jephcott (Stanford, CA: Stanford University Press, 2002).
66. For a similar argument as the one developed here on the social performativity of technical standards, see Amy E. Slaton, *Reinforced Concrete and the Modernization of American Building, 1900–1930* (Baltimore, MD: Johns Hopkins University Press, 2001).
67. Dewey, *Reconstruction of Philosophy*, 197–98.

CHAPTER FIVE

PLYSCRAPERS, GLUESCRAPERS, AND MOTHER NATURE'S FINGERPRINTS

Scott Gabriel Knowles and Jose Torero

In 2013, Canadian architect Michael Green recorded a TED talk extolling, with great lyricism and emotion, the virtues of tall timber buildings: "I've never seen anybody walk into one of my buildings and hug a steel or concrete column," Green told the audience, "but I've actually seen that happen in a wood building. I've actually seen how people touch the wood. And I think there's a reason for it. Just like snowflakes, no two pieces of wood can ever be the same anywhere on Earth. That's a wonderful thing."[1]

At this writing, Green is the designer of the tallest wooden structure in the United States. Given that this structure (an office building in Minneapolis erected in 2016) is only seven stories tall, Green's praise for timber high-rises may strike us as extravagant. His TED talk (now viewed over 1.25 million times) makes a case for tall timber construction as no less than a means of establishing life and work in unique, organic settings, surrounded by "Mother

Nature's fingerprints." But a second argument in the talk, no less passionate, begins to clarify why so many timber-construction advocates trust that their enthusiasm will find receptive audiences. Here, Green moves from aesthetics to sustainability, and we see that the material resonates powerfully with the combined ethical and economic demands associated with environmental awareness, particularly in cities where the great majority of Earth's inhabitants already live or will move to over the next generation. Citing a statistic that three billion people will need a new home by 2040, Green explained to his TED audience that "cities are built in . . . steel and concrete, and they're great materials, they're the materials of the last century, but they're also materials of very high energy and very high greenhouse gas emissions in their process . . . almost half of our greenhouse gases are related to the building industry." How, then, to deal with the world's housing needs and the staggering environmental effects of construction simultaneously? Timber: sustainably farmed trees—natural carbon sinks that they are—provide for many audiences the promise of a sustainable and aesthetically enriched built environment for the twenty-first century.

Green is hardly naïve about the very real impediments to tall timber construction, circulating alongside advocacy since tall wooden buildings emerged as a global architectural focus in the early 2000s. First and foremost is the historically resonant concern over fire—the primary reason that wooden buildings of any significant height remain mostly banned in cities around the world. The control of wood as a building material in densely populated urban spaces marks out a central regulatory focus of the industrial-age city. Indeed, the rise of materials like concrete and structural steel in the late nineteenth and early twentieth centuries succeeded in large part due to their fire resistance, structural integrity, and adaptability to being mass produced. By the 1920s, a wooden skyline was a marker of a city lost to the past—premodern in style and in technology—a relic of a bygone city.[2] Modern zoning, building, and fire safety codes emerged at just this moment in time—alongside standard-setting and testing bodies like the National Fire Protection Association and Underwriters Laboratories.[3] To counteract this fear of fire, irrational (but demonstrable) in Green's telling, he points out the fallacy in thinking that massive pieces of structural timber readily catch fire. Echoing the apostles of "slow burning mill construction" of the nineteenth century, Green argues that under normal conditions, heavy "mass timber" buildings should not burn more easily than any other type of building. To Green, perception is the key to progressive technological change, and the architecture of timber needs an "Eiffel Tower moment"—a demonstration of its safety and of its reliability among the other giants on the skyline. Timber is, Green concludes in his TED talk, "the most technologically advanced material I can build with; it just so happens that Mother Nature holds the patent, and we don't really feel comfortable with it, but that's the way it should be, nature's fingerprints in the built environment."[4]

Green's association of patents not just with nature, but with the eternal beneficence of Mother Nature, is striking. The poetics of nature notwithstanding, Green's framing resonates with a perennial aesthetic concern to modern Euro-American architects and engineers, and one that is essential to any critical history of industrial materials: the intermingling of the natural and the manufactured, and the historical desire to hold together the most satisfying and profitable properties of each category as buildings are designed. As a wholesome material, wood may very well top all others in an ecoconscious cultural milieu such as Green's, but as an industrial material, it faces severe challenges, left in the past by the interrelated cultural priorities of modernism and mass production. As historian Gregory K. Dreicer has explained, even though the mass production of structural lumber was as industrial as any other type of production in the nineteenth century, the visibly organic qualities of wood and its abundance in preindustrial nations led to its wide designation as a premodern technology (if it was seen as a technology at all). Dreicer notes crucially that technical and design expertise was itself partially defined by a choice of "scientific" materials in the late nineteenth century. "The professionalization of engineering coincided with the development of structural wrought iron."[5]

The recent history of engineered wood products, with the use of wood deployed as a strategy of turning "back to the natural," opens this discussion anew and forms the focus of this chapter—a dialogue on the social and cultural processes required to manufacture a new understanding of a supposedly natural product. Without modifications to the industrial processes and regulatory systems for building materials in place worldwide, tall timber will be relegated to tree houses and fail to compete with steel or concrete. If it succeeds as a construction material, it will not simply be a victory of aesthetic judgment. Nor will the take up of timber for high buildings simply represent an assertion of public confidence in a new (very old) material. That success will also come as a result of industrial inventiveness, an intrusion of a new or newly promoted material into a crowded marketplace, and the success of engineered wood in satisfying regulators and safety experts worldwide that it is worthy of inclusion in the collection of approved industrial construction materials.

The "tall timber industrial complex" selectively invokes the natural, sustainable, and safety-engineered characteristics of tall wood structures, and in doing so also reveals the historically contingent relations among regulatory and market actors. No trait associated with a given material is ever understood by producers or users without reference to existing alternative technologies and their implications. Nor are these implications unchanging over time. Across the spectrum of materials histories, we find outcomes that defy teleology: plastic, for example, has gone in the eyes of many from a futurism of unlimited shapes and possibilities to cheap, poisonous, and ocean-destroying in a few decades. From the days when critics questioned John Augustus Roebling's use of wire rope to hold up the Brooklyn Bridge (it doesn't look substantial!) to the disaster investigations following the collapse of the World Trade Center, "fire proof" structural metals have also faced scrutiny over their qualities of stability and fire resistance. The argument for wood as a material both industrial *and* environmentally nourishing—a palliative for climate change—is one we should deconstruct carefully with the knowledge that this rhetoric will be playing against an interrelated set of debates over concrete, steel, glass, plastics, and myriad hybrid combinations of building materials packaged as the "future" of the built environment. We should be attentive to the fact that defining wood or any other material as safe and desirable for high-rise construction is an ongoing task in which public perception, architectural rhetoric, the pressures of the marketplace, and the technical concerns of regulators are constantly interacting, each sector forwarding its own discrete claims and responding to the claims of others as the features associated with reliable building materials by different actors intersect, clash and reform.[6]

ENGINEERING WOOD: MAKING AN OLD MATERIAL NEW

Mother Nature might hold the patent for wood in the form of trees, in Green's lyrical vision, but the technologies that make it possible to produce mass timber panels for high-rise construction are patented and produced by large companies that have worked since the 1990s to bring wood products into widespread use for construction. These include, prominently, the Swedish-Finnish company Stora Enso and the Austrian firms KLH, Binderholz, Mayr Melnhof, and Hasslacher. These firms' commitment to the technology came only a few years after the first report of the International Panel on Climate Change (IPCC) and the awakening of a global market in automobiles, appliances, and architecture for green and sustainable products—a trend symbolized by the emergence of the United States Green Building Council's Leadership in Energy and Environmental Design (LEED) building certification in 1993.[7]

The central invention in engineered wood products is cross-laminated timber, or CLT. The American Wood Council defines CLT as:

> an engineered wood building system designed to complement light- and heavy-timber framing options. Part of a new product category known as massive (or "mass") timber, it is made from several layers of lumber board, stacked crosswise (usually at 90 degree angles) and glued together on their wide faces. This cross lamination provides dimensional stability, strength and rigidity, which is what makes CLT a viable alternative to concrete, masonry and steel in many applications.... [Advantages include] just-in-time fabrication and job site delivery, speed and efficiency in construction, reduced job site noise and on site labor force, substitution of high embodied materials with a renewable resource that sequesters carbon, and creating a living or work space that has the aesthetics of exposed wood.[8]

CLT is a cousin to plywood, particleboard, and other glued-together wood amalgamations. However, CLT marks the transformation of lightweight wood-framing materials, either in the form of composites or cut lumber, customarily seen at construction sites, to a material destined for far more ambitious structural incorporation: this is an engineered wood that its manufacturers hold to be every bit as innovative and reliable as structural steel or concrete.

In addition to CLT, the engineered wood marketplace also notably includes nail-laminated timber (NLT), a product similar to CLT but joined with nails rather than with glues or adhesives. Also included in this product line is glulam, or glued laminated timber; again, a product composed of glued wood pieces, but in this case with the grain running in the same direction. Additionally, the market now offers laminated vernier lumber (LVL), a product in which small wood slices are glued together. LVL is primarily a composite, involving almost as much glue as wood—the opposite of CLT, which constitutes mostly wood with very little glue. Each of these products is highly refined to remove the normal "imperfections" of wood in its natural state, so as to avoid the weaknesses that would be present at knots and other irregular spots in a tree felled directly from the forest. Thus the products' manufactured or engineered character derives not simply from changing the form taken by wood from tree to building element, but also from valuating particular features of the "natural" material differently, and, in fact, rejecting altogether *some* nature (here, knots) as ill-suited to the new application of wood in tall structures. Another natural feature of wood—flammability—also reemerges as an imperfection, but also, like knots, is not a deal breaker for pro-wood architects and producers. As always, industry's blend of interventionist impulses and deference to (what is seen as) preexisting nature is contingent on social and cultural context.

All of these wood-centered materials have been brought out into a climate of growing environmental sensibilities, presenting to producers a wide range of reasons, and options, for expressing their commitment to such priorities. The 1990s saw the arrival of the first IPCC Report, along with the mainstreaming of "green design." For tall buildings—an entrenched priority of urban architects and developers globally by the mid-twentieth century—emergent challenges involved reducing a building's carbon footprint and energy consumption during the structure's entire lifecycle. This meant a new focus on materials, not only in terms of their configurations and performance once constructed but also throughout the manufacturing processes of building materials. Concrete and steel have tremendous "embodied energy"—that is, the total amount of energy consumed from raw materials extraction through finished-materials production—and, therefore, they are not considered to be good candidates when it comes to achieving environmental sustainability. Common estimates hold that concrete manufacturing comprises roughly 5 percent of global carbon emissions, and steel production contributes about 7 percent. Taken together, the modern skyscraper wonder-materials of concrete and steel are major contributors to climate change.[9]

By the mid-2000s, government subsidies across Europe and North America augmented a very strong push from the design community and environmental interests to change the carbon footprint of the built environment. Specifically, incentives in wood-producing countries like Norway, Austria, and Canada went into the timber industry to start creating so-called clean plantations that allowed the responsible production of massive quantities of construction timber.[10] Once in place, such large-scale timber operations made it possible for countries to aspire to actually remove more carbon from the atmosphere than they were responsible for creating—a situation known as having a "negative carbon footprint." With these aspirations have come all sorts of investments and incentives to try to create commercially viable engineered timber products.

The manufacturers taking up these enthusiasms have entered the marketplace via two paths. One path has retained the essential flammability of wood, offering a lightweight construction system where the timber is encapsulated in noncombustible materials that will ultimately act as a protective layer. The other path involves the aforementioned massive timber elements. Because they are very significant in size, it is possible to bring the massive timber elements into the market without any integral fire protection. In name, this is still wood, but it will not burn in the same way wood burns when in the form of lightweight frame structures. In this case, fire safety is predicated on self-extinction. Engineered wood will ignite in a fire but will not continue to burn after the furnishings contained in the burning structure have themselves been consumed. Thus, massive timber elements will essentially behave as would steel or concrete beams, columns, ceiling, or floors.

That behavioral binary—of light versus heavy timber construction encountering flame—represents two broadly competitive market segments, and the claims of CLT makers and related producers restate that difference often. But the binary hides significant cultural priorities with which materials manufacturers contend. Massive processed-timber elements do not burn with the speed of lighter structural elements made of wood, certainly, but they do burn. In order to be understood by designers, builders, investors, and occupants as safe, CLT buildings must meet longstanding ideas of what structures are safe for occupants and property. What follows are descriptions of three areas in which the proponents of massive timber construction—among them engineers, suppliers, architects, planners, and building owners—now exert themselves as they strive to rehabilitate the reputation of wooden buildings in line with their environmental, commercial, or design aspirations.

PROJECT #1: FACING THE CODE MAKERS AND THE REGULATORS

In 2015, CLT was approved in the International Building Code (IBC) for the first time for use as a structural building material in the United States. The IBC is a model building code published every three years by the International Code Council (ICC). The ICC serves as the central standard-setting body for building and structural safety standards primarily in the United States (despite its perhaps misleading name), having been established in 1994 through a combination of the various regional building code-setting bodies across the United States. Unlike countries such as Japan or the United Kingdom that use a more governmental code-setting process, in the United States, the ICC produces its standards through a cyclical series of meetings and debates among materials manufacturers, code-enforcement officials in government, and safety advocates. A two-thirds vote of ICC members is required for a code change. The result is a quasidemocratic, private system that states and municipalities adopt in total or in part depending on local needs and politics. [11]

One year after tall timber appeared in the IBC as an approved material, Michael Green's timber, technology, and transit (T3) building went up in Minneapolis; rising seventy-six feet, it was one of only a handful of wood buildings globally rising above six stories. The seven-story T3 used NLT (again, small boards stacked together and nailed into wall panels rather than glued) and was clad in weathering steel. The architectural press swooned, pointing out the rapid pace of construction, with one hundred and eighty thousand square feet of timber framing completed in just ten weeks; the structure's lightness; and the fact that the wood was harvested from Pacific Northwest forests ravaged by mountain pine beetles.[12] The result, according to the *Architect's Newspaper,* "is a simple massing with an airy brightness, thanks to the exposed wood. . . . In addition to being made from sustainable lumber, which is less energy-intensive to extract, the building will sequester about 3,200 tons of carbon."[13]

The T3 may not have been exactly the dramatic proof-of-concept episode that Michael Green was waiting for, but with the Treet building (160 feet) in Bergen, Norway, having opened the previous year, and the Brock Commons in Vancouver (174 feet) opening just a few months later, the T3 did seem to arrive in something of a global Eiffel Tower moment. This is especially true looking forward just a couple of years to the erection of the Mjosa Tower (240 feet) presently under construction in Brumunddal, Norway, and to the River Beech Tower (684 feet) proposed for Chicago. There is palpable global enthusiasm for timber skyscrapers—structures of manufactured wood and glue, beautiful to many, and cautiously permitted by the building code and fire service authorities responsible for public safety.

The combination of technical and economic advantages offered by the massive timber techniques, with their balm for environmental worry, is powerful, but for the historian, these advances also signal a stunning pivot by influential designers and investors away from technological trends once thought by many in those communities to be immovable: the reliance of builders on steel and concrete. But those trends must also be understood as having derived from deep economic, technological, and cultural commitments: It is not at all surprising that the concrete and steel industries have responded with attacks on CLT, lobbying against ICC code changes and against appropriation of US federal research funds for tall timber construction as enacted by the Timber Innovation Act of 2017. The National Stone, Sand, and Gravel Association, for example, has expressed its disappointment in "members of Congress foregoing marketplace fairness by using federal funding to show preference to one building material over another."[14]

The anti-CLT lobbying and public relations strategy is perhaps most clearly explained in the Ready Mix Concrete Association's "Build With Strength" website. Filled with statistics regarding concrete's resilience in the face of fire and other disasters, the website also includes testimonials from firefighters and a helpful map of major, recent American building fires. "The . . . spate of fires in low- and mid-rise structures throughout the country," according to the concrete industry, "is raising questions and concerns about the safety of wood-built buildings. It's clear that codes and inspections are failing to keep residents and communities safe. It's time for builders, contractors, developers, first responders and residents to come together to create new solutions that embrace noncombustible materials like steel and concrete."[15] To maintain their market position, concrete and steel industry groups are unlikely ever to agree that engineered wood can compete with their materials in the realm of fire safety. Although a great many iterations of concrete and steel themselves exist, possibly with greater and lesser risk of destruction by fire, nonwood industries will likely only ever function as antiwood forces. The commercial interests of concrete and steel sectors may demand such doubts about new engagements with wood.

That environment of competing business interests is one reason that code enforcement is neither automatic nor easy; there is no practical understanding of building materials that is not embedded in actors' subjectivities regarding markets, appropriate civil conduct, and legitimate science. In the United States, these are historically inseparable subjectivities. For code officials, the first and more traditional mode of their work involves a "prescriptive" code model. In this approach, the building codes of a city are enforced by code officials who expect builders to rigorously follow prescribed standards. In recent decades, another mode has become more frequently allowed, which is a performance-based system. Here, according to the NFPA, "the designer must demonstrate that the alternative solution meets the previously listed goals and performance requirements to be considered as an alternative equivalent to the prescriptive requirement." Because of its history as a known combustible material—one that facilitated city-leveling conflagrations for centuries before concrete and steel came into common usage—wood has been consistently and heavily regulated in modern building codes.[16] As such, tall timber buildings will either rely on wholesale redrafting of municipal building codes, or they will need to be expressly permitted under allowances for performance-based designs.

With wood absent from the list of safe tall-building materials for so long, one might easily imagine code officials presented with designs for tall timber buildings asking, "We didn't accept it yesterday, why would we accept it today?" But again, safety expectations and standards are mutable and culturally determined. Builders have effectively responded to this sort of endemic doubt with a combination of demonstrations that give confidence to the code authority, and a great many concessions to concerned observers. Take, for example, the Brock Commons building at the University of British Columbia in Vancouver, an eighteen-story (174 feet) timber building designed by Acton Ostry Architects. Architects provided the building with a concrete core because they could not convince nervous officials or developers to allow a timber core. This is an enormous commercial concession, as it effectively it turns a tall timber building into a concrete building. In so doing, any fiscal gain that would have come from quick wooden construction is lost. Furthermore, to make the building possible, the provincial government of British Columbia had to pass new regulations allowing Brock Commons to exceed timber structure height limits, with the proviso that the structure comply with rigorous fire standards.[17] To meet these demands, all of the CLT and glulam components in the project have been enhanced by complete encapsulation, with three to four layers of fire-rated Type X gypsum board. This has resulted in a building that is even more resistant to fire than an equivalent concrete or steel tower.

Such caution, despite rhetoric from engineered wood manufacturers that might seem to indicate an unchallenged growing reliance on the material for high-rise structures, defines the tall timber discussion at present. Enormous unknowns remain in the eyes of many influential stakeholders. For example, how will exposed junction points within buildings perform in fire? According to a report by the Inland Marine Underwriters Association: "Performance of timber connections exposed to fire can be quite complex based on type of fastener, geometry of connection, [and] different failure modes. Most building codes, including the IBC, do not provide specific fire design methodology for determining the fire performance of timber connections."[18] Gypsum wallboard is required to protect connections and fasteners. Also of concern: How will the buildings perform after a fire has occurred? Can a structure be repaired, repainted, and used safely once the CLT panels have been charred and exposed to water? Such performance specifics, despite assurances from the engineered wood industry, remain unresolved among those whose confidence a widened deployment of timber would require.

PROJECT #2: WOOD BEYOND FIRE

In a sense, the debates shaping scientifically informed code and regulatory decisions can be seen as discretionary, as issues that need only halt planning for those paralyzed by fear of what has not yet happened. Consider that while a timber building stands and it has not burned down, there is in fact a form of proof that timber is safe. The tall-building designers and firms that believe in wood have every reason to keep working on engineering solutions to nonfire related problems, and these solutions continue to mount up as the timber advocates stick with their cause and investors stand loyally by. As fire protection experts and laypeople alike know intuitively, the problem with timber is that timber burns. But architects, planners, and investors work with what must feel to them like additional knowledge, stepping deliberately into the realm of possibility: When (not if) the fire protection problem is solved to the satisfaction of enough of the stakeholders, the tall timber building actually offers advantages over more familiar (and familiarly safe) materials.[19]

Once again, the type of tall timber building under discussion makes a major difference in actors' articulation of such possibilities. For example, in the case of lightweight timber construction, engineers know that a height of about six stories is about as high as one can build, because structurally it's very difficult to create a timber frame that can go higher than that. If a building doesn't weigh enough to offset physical conditions resulting from its height, such as increased sway or wind load, it doesn't anchor itself properly to the soil. Engineers have for many decades confidently calculated wind loads that could topple such buildings, a condition called *uplift*—the risk of which requires complex foundation arrangements to keep the building in place.

A heavy timber building by contrast is very well anchored to the soil, so the structure can endure high wind loads without danger. For heavier timber construction, the sweet spot for height is mostly seen to be at a range between eight and thirty stories. In fact, in that height range many engineers are confident that timber will improve upon any other construction material in terms of overall cost because builders gain enormously on the speed of construction. All of the wood materials in such buildings are prefabricated. During fabrication, weak sectors of the timber (nodes) can be removed and the pieces can be glued to form structural elements that are all of very similar strength and qualities. The result is that very tight tolerances for strength and stiffness can be kept with CLT and similar products, leading to reliable structural elements readily achieved. At the same time, the weight of the timber elements is very small relative to the capacities of customary building processes (picture the tremendous systems alternatively required to move steel girders into place). Builders can thus assemble the tall timber buildings at a speed that seems incredible to firm owners, building owners, and investors accustomed to those traditional materials, and even if the material itself is still slightly more expensive than older options, concrete and steel cannot compete in overall cost with the gains that timber offers.

In terms of assembling the building, massive timber has other advantages as well. Engineers agree that the prefabricated steel stud systems on which older designs rely have great difficulty delivering tight tolerances and repeatability, leading to deep challenges in creating solid and safe finished structures. The problem with the steel studs is that in the holeless tempering process to which they are subjected, the tolerances are very poor—the studs deform. Steel studs are in fact of such light weight that if a wall using steel studs is assembled in a factory, by the time it arrives at the construction site and is put in place, deformations, gaps, and holes may well be present. In this instance, the completed building loses compartmentalization, a core principle of fire protection. By contrast, the stiffness of a thin wall created with timber will be so much greater than one built with steel studs, and therefore the deformation so much less that the building as a whole attains much better tolerances. When timber is used, the overall system retains its shape, is much easier to assemble, and overall could provide a much safer system from a fire protection engineering perspective. Each of these technological considerations runs against the received wisdom that wood is a preindustrial material and is therefore inferior to steel. Neither attribution is unconnected to the interests of its claimants. In this sense, the tall timber debate is as much about circulating perceptions of the material (regarding whether or not wood is safe and reliable) as it is about the technical possibilities that wood enables.

These technological conditions are just some of many involved in the construction of safe buildings, and not surprisingly, perhaps, the fire protection engineering field is presently split over the rise of engineered wood and tall timber construction. The opportunities, like the enthusiasm of the advocates, seem boundless in the eyes of those for whom engineering may be expected to provide safety (not just engineers, but their clients and others in the sectors where they ply their expertise). Yet, the pressure on these safety experts is great. There are more proactive sectors of the construction industry (including fire protection engineering firms like Arup) that are trying to learn as rapidly as possible about the risk of fire in tall timber. They are eager to identify issues and solve them in what they believe is the most responsible manner possible.

These private-sector actors, at present, have fewer resources than they would like for research, and cannot move as fast as the timber industry itself would wish them to move. This leaves many timber industry actors with what they see as a reputational problem in need of a fast solution. Their response to this pressure is to conduct complex, expensive, and large demonstration tests that have, as a single intention, the enacting of fire-safe performance to the public. In many cases, because of the novelty of the circumstances, the fire service and authorities do not yet have the knowledge to detect in these instances the nuances, the complexity, and the risks at levels found in testing regimes for more established industrial materials and processes. The performance of such tests nonetheless proceeds.[20] It can be argued that these "demonstration tests" (variants of Green's Eiffel Tower moments) add nothing to the understanding of the problem from the vantage point of safety, and instead create further confusion by introducing what selected stakeholders see as definitive findings on which to base best practices. Indeed, many of the demonstration tests to date have shown behaviors opposite to what they are ostensibly trying to demonstrate, and the utility of the tests' conclusions for safety is only accepted through ignoring considerable uncertainties. This is assuredly not a new pattern in the creation and marketing of novel industrial materials. Ironically, concrete makers seeking to establish a solid reputation for their material amid strong market pressures first took up this same performative work and staged public fire and strength tests of their products over a hundred years ago.[21]

PROJECT #3: BUILDING WITHOUT BURNING

If we are to think about materials embodying values—regarding the natural and manufactured, the safe and unsafe—we need to think about buildings as more than places where certain materials end their lives. That is, we must consider that wood, or steel, or concrete, bring buildings into being. The complexities of engineered wood products, manufactured in factories and subjected to regulatory regimes at local, federal, and even international scales, force us to consider the character of these products as such: Their flammability, weight, cost, conditions of transport or use, etc. At the same time, we must be aware that when we consider the use of timber building elements, we are talking about a sequence of events with some interdependences, a continuum that runs from the forest to the engineered wood factory, to the construction site, to the finished building, and beyond, throughout the life course of the building and its maintenance regime.

For one thing, engineered wood products are complex composites in which different polymers, wood (natural) and glue (industrialized), are mixed together to achieve a certain level of performance. This performance has been studied by producers and regulators in terms of structural integrity, acoustics, and durability.

Indeed, many of what materials producers, building designers, and regulators have thus far taken to be the most complex *structural* problems have been solved. When it comes to fire, the same scale and scope of effort have not been made as of yet, and because of the historical trajectory away from tall timber construction, there is no dedicated research yet that allows engineers and builders to understand with real depth the behavior of these complex composites in fire. Furthermore, as opposed to other performance criteria (like those established for completed structures), there is no study of the coupled behaviors of the hazard and the system. For example, when addressing wind loads and uplift, the timber system has to show performance against a predefined wind load that is not affected by the choice of material; all materials must meet that wind load or the building will fail. When it comes to fire, materials behave in profoundly different ways in different combinations, and simply asking for "fire safety" captures few of these complexities. For example: VLT can delaminate as it degrades, falling apart, in which case pieces of timber can fall and feed the fire, thus extending the duration of the fire. If such delamination continues, then the falling timber will feed the fire until the building is consumed. Thus the construction material can be seen to couple with the hazard in a highly complex way. The timber industry on the whole, in their rush to enter the tall-building market, has not yet addressed the risks that come with the complexity of the systems they are proposing.

For all the debate about the risks associated with occupied timber structures, the more pernicious fire threat for tall timber construction is not that associated with a finished building, but that residing in the poorly regulated phase of construction itself. Here is where the dangers of wood perhaps fulfill our worst nightmares. Tall timber construction fires spread rapidly to involve the entire construction site. Such fires have proven dangerous for fire services. Recent fires of unfinished tall timber buildings at the University of Nottingham, and in apartment complex projects in Los Angeles and Edgewater, New Jersey, demonstrate that the unfinished, not-yet-encapsulated timber building can burn ferociously, with fires reminiscent of nineteenth-century conflagrations.[22] In effect, timber construction sites are to date being managed in the same way one might manage a concrete or steel site, entirely inadequate from the vantage point of construction personnel, community, and firefighter safety.

To eliminate fires in construction will require reformulating building practices and thus yet more research on the deployment of wood for large-scale construction, here analyzing the practices of the construction site and how to manage them in a fire-safe manner. It can be surmised that because such fires are relatively inexpensive from a fire insurance perspective (the buildings aren't complete yet), and the risks are predominantly borne by firefighters (not tenants), that there is a low incentive to address the construction fire problem. This clear pattern of precompletion risks—destroying investments long before buildings are ready to use—tends to evade examination by those concerned with the safety of completed structures. Perhaps this neglect arises because the loss of life associated with such fires is less easily pictured, and any loss of property is largely confined to commercial firms and investors bringing the building into existence, not with the general public. Yet, the very difficulty of conceptualizing risk and damage in this case tells us something about the economic and political priorities of contemporary commerce and development. What's more, and perhaps more fundamentally for our historical study of materials, our attention to these many phases of timber construction reveals the astonishingly complex encounters of routine building processes and building materials in various stages of preparation and assembly. Any semblance of ease or clarity in establishing the safety of timber construction dissolves under this scrutiny.

CONCLUSION

Standards of safety, of sustainability, and of aesthetic value are, without question, in dispute for what critics have termed the "plyscraper," or the "gluescraper"—standards for a new material that is also, arguably, classifiable as one of the oldest materials one might imagine, carrying the anxieties about fire risk that timber has always borne as a building material. The complex twenty-first century recipe of safety in the short term for building users, and longer-term well-being in the face of devastating climate change, blends in the tall timber building with other elements: the perpetual building-owner concerns of architectural prestige, return on investment, and commercial expedience. When Michael Green shares his effusive love of nature, expressed in the timber skyline, we must also be aware that this emotion moves amidst a complex choreography of fire experts, materials engineers, regulatory agencies, and insurance brokers for whom such love is a more daunting prospect (and whose voices may be less easily heard than Green's in the public square).

The recipe of technological innovation and natural wholesomeness that Green celebrates in discussing massive timber building elements is perhaps more telling about American industrial values than he realizes. The tall wooden building of the twenty-first century not only uses an "old" material, it must also contend with old anxieties about that material and the established professional and bureaucratic institutions that have taken building safety as their purview for generations. We have no reason to doubt the sincerity of those taken with the idea of timber skylines: designers, environmental activists, and cost-conscious building firms or investors. But one cannot honestly promote the novel characteristics of CLT and related products without confronting concerns and risks that have been with us since antiquity.

There is one more way in which the tall timber building is not new. Since the first Euro-American efforts to commercialize the manufacture of building materials, in the later nineteenth century, claims of technological innovation and aesthetic achievement have built on one another. The two virtues have by now long been linked in modernizing cities, globally, and the take-up of new building materials and meaningful cultural contributions feel virtually inseparable. Surely the wood skyscrapers enact the same twinned functions. But as well, if green values are today truly, sturdily joining them as a third signal of cultural and civic virtue, as a deep commitment of industrialized societies, it may be the emergence of the tall timber building that demonstrates this expansion.

NOTES

1. Michael Green, "Why We Should Build Wooden Skyscrapers," TED Talk, 2013, https://www.ted.com/talks/michael_green_why_we_should_build_wooden_skyscrapers#t-670252.
2. Amy E. Slaton, *Reinforced Concrete and the Modernization of American Building, 1900–1930* (Baltimore, MD: Johns Hopkins University Press, 2001); Reyner Banham, *A Concrete Atlantis* (Cambridge, MA: MIT Press, 1986).
3. On fire safety and the history of the built environment, see Scott Gabriel Knowles, *The Disaster Experts: Mastering Risk in Modern America* (Philadelphia: University of Pennsylvania Press, 2011). See also Mark Tebeau, *Eating Smoke: Fire in Urban America, 1800–1950* (Baltimore, MD: Johns Hopkins University Press, 2003); and Carol Willis, *Form Follows Finance: Skyscrapers and Skylines in New York and Chicago* (Princeton, NJ: Princeton University Press, 1995).
4. Green, "Why We Should Build." For a sense of the multiple, culturally specific meanings historically accorded to wood as an alternative to metal by building designers, and a discussion of aesthetic impressions sought by those using each material, see Darin Hayton's chapter in this volume.
5. Gregory K. Dreicer, "Building Myths: The 'Evolution' from Wood to Iron in the Construction of Bridges and Nations," *Perspecta* 31 (2000): 130–40, 138.
6. For a discussion of materials and historical contingency, see Jeffrey Meikle, *American Plastic: A Cultural History* (New Brunswick, NJ: Rutgers University Press, 1997); and Miles Orvell, *The Real Thing: Imitation and Authenticity in American Culture, 1880–1940* (Chapel Hill: University of North Carolina Press, 1989).

7. United States Green Building Council, "USGBC History," accessed June 17, 2019, https://qas.usgbc.org/about/history.
8. Layne Evans, "Cross Laminated Timber: Taking Wood to the Next Level" (Leesburg, VA: American Wood Council, n.d.), 1; http://www.awc.org/pdf/education/mat/ReThinkMag-MAT240A-CLT-131022.pdf; Paul Coats, "Introduction to Mass Timber Construction" (Leesburg, VA: American Wood Council, 2017), 4, https://www.usg.edu/assets/facilities/documents/foc/2_Introduction_to_Mass_Timber_Construction.pdf.
9. Madeleine Rubenstein, "Emissions from the Cement Industry," *State of the Planet*, The Earth Institute, May 9, 2012; http://blogs.ei.columbia.edu/2012/05/09/emissions-from-the-cement-industry/; OECD, "Making Steel More Green: Challenges and Opportunities," Workshop on Green Growth in Shipbuilding, Paris, July 7–8, 2011; https://www.oecd.org/sti/ind/48328101.pdf.
10. Jack Fraser, "Knock on (Engineered) Wood: Pathways to Increased Deployment of Cross-Laminated Timber: A Case Study of the Building Sector in Sweden, Master of Science in Environmental Management and Policy," (MS diss., Lund, Sweden, September 2017).
11. On the International Code Council, see "About the International Code Council," International Code Council, Inc., 2019, https://www.iccsafe.org/about-icc/overview/about-international-code-council/.
12. Jenna McKnight, "Michael Green Completes Largest Mass-Timber Building in United States," Dezeen, December 2, 2016, https://www.dezeen.com/2016/12/02/michael-green-architecture-t3-largest-mass-timber-building-usa-minneapolis-minnesota/.
13. Olivia Martin, "Timber!: Largest Mass Timber Building in U.S. Opens Tomorrow in Minneapolis," *Architect's Newspaper*, November 29, 2016, https://archpaper.com/2016/11/t3-minneapolis-mass-timber-building/#gallery-0-slide-0.
14. National Stone, Sand & Gravel Association, "NSSGA Opposes Timber Building Bill," https://www.nssga.org/nssga-opposes-house-pro-timber-building-bill/.
15. Build with Strength: A Coalition of the National Ready Mixed Concrete Association, "America Is Burning," http://buildwithstrength.com/america-is-burning/.
16. According to the NFPA, "Prescriptive height and area limits for noncombustible materials, such as steel and concrete, are practically unlimited. As a combustible material, prescriptive limits on timber buildings are generally capped at eight stories or less." National Fire Protection Association, "Fire Safety Challenges of Tall Wood Buildings," Fire Protection Research Foundation, 2013, 21.

17. Len Garis and Karin Mark, "Tall Wooden Buildings: Maximizing Their Safety Potential," Fire Engineering, January 1, 2018, http://www.fireengineering.com/articles/print/volume-171/issue-1/features/tall-wood-buildings-maximizing-their-safety-potential.html.
18. Erik G. Olsen, "CLT and Builder's Risk," Inland Marine Underwriters Association, 2017.
19. See Susan Deeny, Rory Hadden, Barbara Lane, and Andrew Lawrence, "Fire Safety Design in Modern Timber Buildings," *The Structural Engineer* (January 2018): 48–53. For fire safety and performance testing, see also Robert Gerard and David Barber, "Fire Safety Challenges of tall Wood Buildings," Fire Protection Research Foundation, National Fire Protection Association, 2013.
20. On the value of testing as a means of forestalling doubt, see Geoffrey C. Bowker, *Science on the Run: Information Management and Industrial Geophysics at Schlumberger, 1920–1940* (Cambridge, MA: MIT Press, 1994).
21. Slaton, *Reinforced-Concrete*, 68–88. For an overview of critical science studies analyses of such public tests, and distinctions between testing and demonstration initiatives associated with public safety, see Harry Collins and Trevor Pinch, *The Golem at Large: What You Should Know about Technology* (Cambridge: Cambridge University Press, 1998), 57–75.
22. "University of Nottingham Laboratory Fire Caused by Electrical Fault, Says Report," BBC News, January 9, 2015; https://www.bbc.com/news/uk-england-nottinghamshire-30751431; "Up In Smoke: When High-Rise Buildings Become Tinder Boxes," The Economist, December 24, 2014, https://www.economist.com/news/2014/12/24/up-in-smoke.

PART III

MATERIALS INTERPRETED, COMMUNITIES DESIGNED

CHAPTER SIX

THE INMATE'S WINDOW

Iron, Innovation, and the Secure Asylum

Darin Hayton

On December 25, 1813, the members of the Building Committee of the Friends' Asylum in Philadelphia distributed the tasks related to building the proposed asylum. Along with determining the best place for a stone quarry, acquiring lime and stone, and finding a stone mason and carpenters to do the work, the committee appointed Samuel P. Griffitts, Joseph M. Paul, Jonathan Evans, and Thomas Wistar to the subcommittee on "the admission of light and air."[1] The Subcommittee on the Admission of Light and Air was responsible for the design, fabrication, sizing, and placement of all exterior windows. Over the next three years, the committee's purview expanded to include the design and construction of interior transom windows into patients' bedrooms and even the locks on patients' doors. What linked window sashes to door locks as committee responsibilities was a Quaker understanding of "insanity" as caused by socioenvironmental conditions. Treating persons seen to be insane thus depended on creating and maintaining restorative socioenvironmental conditions. For Philadelphia Quakers, such restorative conditions were produced in the combination of domestic aesthetics with ideals of security for both the insane and for their caretakers. The Quaker domestic aesthetics were secure, and asylum security was aesthetic. When designing and fabricating window sashes and door locks, the Philadelphia Quakers transformed iron, a traditional asylum building material, by deploying it in new forms in service to their desire to construct a pleasing and secure asylum. Their recognition that window sashes and door locks were problems to be solved and their sustained efforts to design, evaluate, and fabricate material solutions—to comprehend both problems and solutions as ethical, aesthetic, technical, and even tactile challenges—derived from their understanding of insanity and evinced their new ethos of care.[2]

When members of the Quaker community in Philadelphia decided to erect an asylum for "Persons Deprived of the Use of Their Reason," they emphasized their innovation in the treatment of the insane. They claimed to reject the earlier regime of physical coercion reliant on therapeutics of restraint, corporal punishment, and threats of deprivation that sought to compel patients to act rationally. They replaced that regime with therapeutics of kindness, a "moral treatment."[3] The asylum-founding Quakers believed that the frenzies and violent behavior of the mad were products of the brutal treatment these unfortunate persons had endured, not justifications for such treatment. The Quakers' moral treatment of the insane denied organic or physical causes and focused instead on the emotional and rational causes of madness. The Quaker ethos of care emphasized the patients' humanity and ascribed to them a fundamental role in recovering their reason; an agency that disrupted longstanding ideas of the mad as being without the capacity to act in meaningful ways. Moral treatment was built around nonmedical therapeutics that managed all aspects of a mad person's environment without the appearance of coercion or restraint.

The Quakers sought to create means of confinement without the appurtenances of imprisonment. The asylum should be domestic rather than punitive and should embody "not the idea of a prison, but rather that of a large rural farm" while ensuring safety and security for all members of the family.[4] In the Quaker asylum, appearance and affect were fundamental features of treatment.

In their efforts to design and build an asylum, Quakers read traditional building materials through their new ethos of care. Their belief in psychosomatic causes of madness replaced ideas about somatic causes. Quakers brought their new understanding of insanity to the design and construction of the asylum and, consequently, considered the role that materials played in patient care and imbued construction materials with real agency. In their new asylum, construction materials in any form, whether iron window sashes or door locks, glass windowpanes, or even stone steps, compelled and impeded particular actions, and regulated behavior in ways that enacted social relations and values.[5] Why not, we might ask, a "Committee on Doors and Windows" rather than one on "Light and Air"? Because the encounter of the Quaker-built structure with nature's light or atmosphere was one of materials engaging with materials, of physical, sensory arrangements that enlisted the nonhuman very broadly imagined (glass and light, walls and fresh air, metal and noise) in the Quakers' humanistic, experiential project.

The asylum planners' new affective priorities transformed even iron from a material of confinement to one that established a sheltering domesticity.[6] In its appearance and tactile characteristics—even the sounds emitted by moving parts—iron itself could provide care.[7] As one contemporary reviewer of psychiatric care practices put it, "The only objection to the use of iron, is the name."[8] In addition to reexamining the raw materials used to construct the asylum, Quakers also reevaluated how those materials were shaped, fabricated, and deployed in the asylum. Their moral treatment of insanity created a set of problems from aspects of everyday life largely invisible to other institution builders—the affective nature of window sashes and mullions, and door locks—and identified possible solutions to those problems, or at least the possibility that those problems could be solved.[9] The Quakers who founded the Friends' Asylum created a system of committees that mapped onto their emergent understanding of patient care. Each committee delineated a particular problem and established the domain of possible solutions. Their efforts are not appropriately disaggregated in retrospect as reflecting either technological or aesthetic priorities, as the Quakers' standards of workable artifact and pleasing or economical design gave meaning to one another.[10] Through material and design choices, work within each committee distinguished between proper and improper solutions to asylum-making problems.

Scholars have analyzed in great detail the architecture and design of nineteenth-century asylums and have shown how the changing ideas about the nature of insanity and emerging treatment orthodoxy shaped the physical spaces both inside and outside the asylum walls. In this new understanding, asylum design itself became an instrument of therapy, assisting patients in developing appropriate habits. Properly considered and executed, design enhanced treatment and the healing process.[11] While these histories have helped elucidate the symbiotic relationship between ideas about madness and asylum architecture, their focus on the design and usually architectural ideals of the historical actors has tended to lend their scholarship an abstract quality. Design in these accounts tends to do little more than reflect the conceptual developments in the history of medicine.[12] By examining window sashes and door locks, we see how values are made manifest in the very materials of the asylum. We see innovation, design, production, and manufacture as undertaken in the service of those values—indeed, as initiatives that are meaningless without the articulation of such values. The Subcommittee on the Admission of Light and Air and more broadly the Building Committee of the Friends'

Asylum struggled to produce sashes and locks that reflected their understanding of madness, but their understanding of madness was, at the same time, shaped by their efforts to design sashes and locks. In this process, aesthetics, security, availability of materials, affordability of products, bureaucracy, and the committee's own expert status all shaped both their understanding of madness and their understanding of the material itself.

THE RETREAT AT YORK AS MODEL

In 1791, Hannah Mills, a young Quaker, died in the York Lunatic Asylum. Visions of Mills chained in a cell, suffering and finally dying alone in the dark and the cold haunted William Tuke, prompting him to organize the local Quaker community to establish their own institution for the insane. Five years later, Tuke's York Retreat began admitting patients and caring for them in an entirely different environment, one shaped at every level by the ideals of moral treatment. This moral treatment rejected chains and manacles, iron bars over the windows, and the appearance of confinement. Instead, as Samuel Tuke, the retreat's chief advocate and publicist, emphasized repeatedly in his published works, the "object of the Retreat, [was] to furnish a comfortable shelter for [an] insane person, as well as to promote their recovery."[13] From the standpoint of helping patients recover, Tuke celebrated comfort as "the highest importance, in a curative point of view."[14]

At the core of the retreat's approach were not simply new ideas about comfort and patient liberty, but new ideas about the nature of madness itself. Although they acknowledged that drugs and somatic medical practices might relieve coincidental symptoms, these reformers doubted medicine's role in bringing about any real cure. In 1817, a review of recent literature on asylums published in the *Edinburgh Review* emphasized medicine's limits in treating the insane: "There is, however, nothing that leads to a belief in the specific efficacy of any particular drugs, or gives reason to suppose that medicine can be useful, otherwise than by relieving or preventing those bodily derangements, which cause or accompany derangement of the mind."[15] More forcefully, they severed the connection between physical coercion and bodily restraint. Prevailing efforts to cure the mad had mistakenly relied on systems of restraint under the guise of safety for both the patient and asylum attendants. Such efforts exacerbated the problem precisely because they drew attention to the patient's outrageous behavior while at the same time neglecting the patient's comfort:

> Many errors in the construction, as well as in the management of asylums for the insane, appear to arise from excessive attention to *safety*. People, in general, have the most erroneous notions of the constantly outrageous behaviour, or malicious dispositions, of deranged persons; and it has, in too many instances, been found convenient to encourage these false sentiments, to apologize for the treatment of the unhappy sufferers, or admit the vicious neglect of their attendants.
>
> In the construction of such places, cure and comfort ought to be as much considered, as security; and, I have no hesitation in declaring, that a system which, by limiting the power of the attendant, obliges him not to neglect his duty, and makes it his interest to obtain the good opinion of those under his care, provides more effectually for the safety of the keeper, as well as of the patient, than all "the apparatus of chains, darkness, and anodynes."[16]

Tuke returned to this idea again and again. The mad fly into rages when handled harshly, when restrained, when confined to cold, dark cells. "Can it be doubted," Tuke asked, "that, in this case, the disease had been greatly exasperated by the mode of management?"[17] Violent behavior arose, in Tuke's view, "from the mode of management" and was "easily excited by improper treatment."[18] Treatment at the new retreat had inverted the prevailing relationship between symptoms and disease. Traditional ideas about madness that considered patients' behavior to arise from the disease justified the use of restraint. Tuke, by contrast, rejected the use of restraints because, in his view, they caused the behavior that was mistaken for a symptom of madness.

The Quaker ethos of care saturated every administrative and material element of the retreat at York with new meaning. William Stark, who wrote a widely cited book on asylum design, praised the retreat, writing that,

> A great deal of delicacy appears in the attentions paid to the smaller feelings of the patients. The iron bars, which guarded the windows, have been withdrawn, and neat iron sashes, having all the appearance of wooden ones, have been substituted in their place; and when I visited them, the managers were occupied in contriving how to get rid of the bolts with which the patients are shut up at night, on account of their harsh ungrateful sound, and of communicating to the asylum somewhat of the air and character of a prison.[19]

His comments highlight the minute details that occupied Tuke and the Quakers who founded the retreat. Iron bars covering windows were removed. In their place, the founders used glazed windows with iron sashes that were painted to look like normal wooden sashes. That the affective impact of sashes—a signal of "normal" windows, and one's residency in a "normal" building—mattered, but that conventional wood sashes did not suffice for security shows us the insufficient "caring" of both familiar iron bars and familiar wooden windows. That the large iron bolts on patients' bedrooms were heard to have a "harsh, unpleasant" voice indicates that the locks were recognized as having a regrettable disposition, although in 1807, when Stark wrote his book, they had not yet been replaced. A better, milder approach to all features of the material surround afforded the patient considerable comfort and individual liberty without sacrificing security.[20]

The retreat at York exercised a direct and immediate influence on Thomas Scattergood and, through him, the Quakers in Philadelphia who founded Friends' Asylum. Scattergood had spent five years traveling through England and Ireland in the 1790s. As early as 1795, he had dined with Tuke in London, and over the next few years, he met with different members of the Tuke family. When he visited York in September 1799, he toured the retreat with Tuke and visited patients there.[21] When Scattergood returned to Philadelphia, he convinced the Quaker community to establish an asylum on the model of the retreat.

IRON SASHES

When Tuke and John Bevans designed the retreat, they overturned what it meant to be a window in an asylum. Given the prevailing ideas about insanity and patients, windows in earlier asylums had been little more than openings in walls, just as they were in prisons. In order to prevent patients from escaping, windows were typically placed high in the wall and secured with iron bars. In some cases, the windows were fitted with shutters that, when closed, prevented light from entering the cell.

For Tuke, the standard asylum window had a number of problems. First, the use of iron bars made the asylum look like a prison and emphasized confinement. Iron bars, like chains, identified the madman and the felon, and privileged confinement over care. The emphasis on confinement must have, according to Tuke, "a gloomy effect on the already depressed mind."[22] The focus on security and safety has exacerbated patients' symptoms and outrageous behavior at the expense of their recovery. Second, placement of the windows high in the wall admitted some light but prevented the patient from seeing outside. In addition to the deleterious effects the gloom had on the patient's mental state, the placement of the window denied the patient the calming effects of a pleasant view. Third, the use of shutters cast the patient's cell into near darkness but did little to keep the cell warm during cold seasons. "Iron bars and shutters, are too often substituted for glazed windows, in rooms appropriated to the insane. The obvious consequence is, that the air, however cold, cannot be kept out of the apartment, without the entire exclusion of light."[23] Tuke believed that insane patients suffered from cold and exposure just as any sane person did, a significant equivalence that would direct his use of resources and choice of design.

Tuke returned to the problem of windows and bars in his *Practical Hints*: "In regard to the *manner of admitting* light; iron bars and shutters, and very frequently, iron bars without shutters, have been substituted for glazed windows in the bed-rooms of the insane."[24] Tuke's understanding of insanity made this model of a window seem barbaric, closely related to the practice of chaining patients to the wall. Tuke's innovation was to see the window itself, to consider what a window was, and to make it part of the treatment. The design, size, placement, and construction of this architectural element were intimately joined to the moral treatment that Tuke believed restored or at least helped restore a patient's sanity. Within this web of meanings, materials held substantial agency. For example, windows must be glazed. At the retreat, "all the rooms, except three which derive their light from an adjoining gallery, have glass windows."[25] Even windows in the patients' bedrooms, which typically had been secured with bars, should be glazed.[26] Moreover, windows should give the institution the appearance of a rural farmhouse.[27] No longer was it sufficient for a window merely to admit light. Instead, windows must be placed at a level that allowed patients to see out, to enjoy a view of the gardens and landscape outside. At the same time, windows must also be sufficiently large for the patients to enjoy while inside. Windows must be cheerful: "The windows [in the day rooms] should be within reach of the patients, so as to afford them the gratification of prospect, security and cheerfulness being combined, by making the sashes of iron, and the panes of small size. . . . The windows of the sleeping rooms, also, should, in general, be of sufficient size to render them cheerful, and should never be without glazing."[28] The windows at the Retreat embodied Tuke's ideas about cheerfulness, comfort, and security but above all a collaborative understanding of care that endowed the ill, at least putatively, with meaningful volition, even in their state of madness.[29] Patients, he claimed, tended to break windows because those windows were placed out of reach. By placing windows within reach, fewer patients would break them.[30] Even the mad might comprehend and rise to the level of reasonable societal expectations.

Or, rise "almost" to that level: this was a calculated idea of volition that did not maximize caregivers' trust in patients. To limit the damage aberrant patients could and, Tuke worried, would do, to minimize the cost of that damage and to prevent their escape while at the same time having a window that looked like a window in a normal house both from the outside and the inside, Tuke designed asylum windows with iron sashes and mullions, along with small standardized panes of glass.[31] Finally, the systems of iron sashes were painted to look as if wood secured the windows. A double sash system allowed for the windows to be opened to admit air or closed to keep the rooms warm. Tuke described the windows installed in the men's day room:

> There are two windows in the room, which afford an agreeable view of the country. They are three feet and a half wide by six feet high, each containing 48 panes of glass, or 24 in each sash. The frames of the sashes are of cast iron, about one inch and a half square; the glass-bars are about five-eights of an inch thick, and each pane of glass is about six inches and a half by seven and a half. Air is admitted through the windows, by placing the upper cast iron sash, not glazed, immediately over the lower one, and hanging a glazed wooden sash, precisely of the same dimensions, on the outside of the iron frame. In this manner the double sash windows, in general, especially in the patients' apartments, are all

effectually secured without an appearance of any thing more than common sashes with small squares.³²

The upper and lower iron sashes of the asylum windows were fixed in place. Although the upper one was divided with mullions, it was unglazed. Outside this fixed sash was a wooden one of the same design and dimensions that was glazed. The wooden sash could be lowered or raised to open or close the window. Tuke had designed a window that addressed Quaker demands for domesticity and comfort and some quantum of respect for the ill, as well as security.

Far from solving definitively the problem of the windows for all those involved in founding subsequent Quaker institutions for the insane, all the attention the retreat paid to fenestrations highlighted the fact that windows were problematic. We are reminded that building is invariably local, a process navigated by the locally influential but also requiring local labor and materials and shaped by local markets. For these reasons, architectural choices are only poorly explained by conventional stylistic timelines. When the Building Committee for the Friends' Asylum met in the fall of 1813, they immediately focused on windows as a feature that demanded consideration. Samuel P. Griffitts and Joseph M. Paul formed a subcommittee "appointed to ascertain the best mode of constructing Iron Sash [sic], and make a report when prepared."³³ Even at this early date, the building committee and, by extension, the board of managers in charge of founding the Friends' Asylum had decided on a number of key features of windows in the asylum. At the most basic level, the committee had determined what a window was. Windows were glazed. They had sashes and could be opened and closed to allow or prevent the flow of fresh air. The sashes would be iron, which satisfied the established aesthetic and security requirements. The committee had also determined that they, the members of the building committee, would oversee and ultimately control the design and construction of the sashes. Two months later, on December 25, 1813, after Griffitts and Paul reported back to the building committee, their subcommittee was expanded, as was its charge: "Samuel P. Griffitts & Jos. M. Paul who were appointed some time since to ascertain the best mode of constructing Iron Sash [sic], reported that they had given attention to the subject; Jonathan Evans and Thomas Wistar are now joined with them, & it is requested that they may take into consideration generally the subject of the admission of light & air into the Building."[34] The building committee had thus created the Subcommittee on the Admission of Light and Air (SALA). This subcommittee was responsible for not only the design and construction of windows, but also the placement and size of windows and the appropriate means of securing them. The subcommittee was further charged with evaluating designs and assessing the feasibility of constructing the windows, as well as identifying craftsmen to cast the iron sashes and negotiating their contracts. In short, the SALA managed all aspects of windows, from their conception and design to the selection of craftsmen, and the evaluation of materials and workmanship through to the installation of the completed window into the walls of the asylum. The subcommittee's members functioned as experts at every stage, from the physical design and architectural placement of this building element to conceptualizing the window's effects on insane patients.

One of the first issues the SALA decided was who would fabricate the sashes. One option was to set up a forge and secure the work of a blacksmith on site. The building committee had established a precedent for carefully controlling such production when it had decided to quarry stone for the building and produce lime on-site. In the case of the ironwork, however, the committee decided that it was better to contract with an established blacksmith, possibly avoiding the cost of an on-site installation that would have likely employed someone hired from the same pool of nearby blacksmiths. In January 1814, the SALA reported back to the building committee their decision not to construct a forge on-site and instead to hire a blacksmith.[35] But money was not clearly the sole determinant of such decisions. Building components such as windows were not mass produced until much later in the century, meaning that even conventional wood sashes were individually fabricated and that commercial suppliers rarely achieved economies of scale. While the United States did not impose protective tariffs on window glass until 1820, and instead imported English panes that were a relative bargain due to high American labor costs, glazed windows likely represented a significant discretionary financial outlay for the asylum builders.[36]

When the board of managers met in March 1814, they formally adopted the use of iron sashes adapted from the retreat at York: "Eighthly. It being judged practicable to have Iron Sashes constructed it was resolved to adopt them on a similar plan to those in use at the 'Retreat' near York in England."[37] As a precedent, York carried multiple sorts of credibility for the Philadelphia Quakers. That same month, the board decided to publish an "Account of York Retreat," which they distributed in the Philadelphia region to promote the Friends' Asylum. Their book included a short account of the Friends' Asylum and listed individual donors as well as monthly meetings that had contributed to the project. Already by 1814, they had secured contributions from twenty-five monthly meetings and nearly three hundred individuals, totaling more than $23,000. Although construction had just begun, the board of managers apologized for the additional cost caused by their decision to extend the front of the main building beyond the front of the wings, claiming that such a modification of the building was necessary to ensure the proper admission of light and air: "A view of the proposed building is prefixed to the present publication. The unavoidable extension of the front, arises from the necessity of affording comfort and convenience to the patients, by procuring a free admission of light and air. This important consideration will lead, in the first instances, to more expense, but we do not doubt will be fully counterbalanced by the advantages resulting from it."[38] Here we catch sight of the great importance of light and air, as the board changed the overall design of the building to ensure the proper placement of windows, and thereby the admission of both light and air into the building, even when that alteration significantly increased construction costs. That same month, the building committee accepted the SALA's recommendation to change the roofline to accommodate the windows.[39] Two months later, they incurred yet further costs when they altered the plans once again and lowered all the windows in the northeast wing by twelve inches.[40] Clearly, the SALA and the building committee saw money not as a boundless resource (otherwise, why go to the trouble to justify expanding budgets?) but as a nonetheless well-spent reserve in the pursuit of the perfect window. Indeed, to decide to spend money on the insane was to calibrate one's level of care for them.

That calibration was not, however, only a matter of scaling up costs. At the meetings of the building committee over the next couple of months, the SALA reported that they were continuing to work on designing the window. At the meeting of the building committee in September, the SALA reported that "castings of sash [sic] had been made by which experiment the practicability of the plan is established."[41] The iron sash included the framing sash as well as the mullions and transoms that separated the individual panes of glass. We see here the SALA acting as expert consumer and ensuring quality control at the level of design and fabrication. Having designed or perhaps overseen the design of an iron sash, the committee then obtained a sample that they could evaluate according to their criteria of aesthetics and security. They also used the sample to assess the availability and affordability of the iron sashes. Here, the SALA acted as quality-control experts. For them it was not sufficient to have designed a window. Only by handling a physical prototype could they determine whether or not the window—its design, material, function, aesthetics, etc.—met their requirements. For the members of the SALA, the tactile experience and the conceptual ideal of windows were inseparable, just as, in the Quakers' ethos of care, the patients' bodily encounters with windows, walls, air, and light, and with the sensation of authentic domesticity, were inseparable. Materiality and affective experience had little meaning without the other.

This was an iterative process for those responsible for producing the optimized asylum. While the subcommittee members' initial "experiment" confirmed the general feasibility of the plan, their initial design apparently needed to be refined. As construction on the building progressed, the SALA continued to work out the design details. In July of 1815, they paid Abraham Lower eight dollars for the final pattern of an iron sash he had submitted to the committee.[42] Lower's design apparently satisfied the SALA and the building committee, which soon agreed that even the transom windows and garret windows should use iron sashes, reversing an earlier decision to secure both with iron bars.[43]

At this point, the SALA had to shift its focus from evaluating samples to finding a blacksmith who could produce the iron sashes to their standards in suitable quantities and according to their scheduling needs. By October 1815, they had entered into a contract with Benjamin Jones to produce the sashes for the 113 large exterior windows and, initially, the iron sashes for twenty of the interior transom windows. Over the next year, Jones produced the sashes for the entire building. The decision to use iron sashes was unquestionably a costly one compared to other options; that is, either conventional iron bars or wooden sashes. Depending on the size of the window, the sashes ranged in weight from 60 pounds to 228 pounds. The sashes for the greatest number of windows (i.e., the exterior windows on both wings) weighed 169 pounds and cost more than ten dollars each. In total, the iron sashes cost more than $1,300.[44]

MORTICE LOCKS

The Quaker ethos of care also induced the building committee to examine locks on patients' doors. While confinement remained a baseline requirement, just as they had reimagined windows in the asylum to marry their concerns about security with their emphasis on domesticity, the asylum planners redesigned the locks used on patients' doors. They rejected on aesthetic grounds the bolts commonly used in asylums, opting instead for a type of mortise lock that they designed and had fabricated expressly for the asylum.

In 1807, when William Starke visited the retreat at York, he noted that the managers there were trying to eliminate the use of bolts on patients' doors. Nearly a decade later, Tuke still worried about the deleterious effects that large bolts had on patients. In his *Practical Hints*, he complained: "The cheerfulness of Asylums has, in too many instances, been sacrificed to security; every door has been guarded by massy bolts."[45] Once again, asylums had wrongly privileged security and confinement over patient care and recovery.

The Building Committee of the Friends' Asylum, however, refused even to consider locking bolts, though initially they did not know what to use instead. Faced with an untenable set of default options—an existing world of asylum locks that failed to express the appropriate ethos—just as they had done for windows, they formed yet another subcommittee, the Committee on Fastenings, which included many of the same members serving on the SALA. Administrative exertions would yet again be called upon to generate design knowledge and muster materials that were Quaker in identity. Among other tasks, the Committee on Fastenings was instructed to determine a suitable lock for patients' doors. Its members surveyed the possible options, and consulted with experts and evaluated their advice before designing their own solution. And just as the SALA had done with the window sashes, the Committee on Fastenings obtained prototypes that they used both to evaluate the lock and to assess the quality of the craftsmanship. In this way, they also determined who could manufacture the locks. Finally, they contracted with craftsmen to produce and deliver the locks according to their schedule.

When initially formed on April 1, 1815, the Committee on

Fastenings wrote to Dr. William Thornton in the United States Patent Office in Washington, DC, to ask if he could recommend any improved locks specifically designed for patients in an asylum: "Samuel P. Griffitts and Roberts Vaux are requested to address a letter to Dr. William Thornton of the Patent Office at Washington City, soliciting from him any information he may possess on the subject of improved locks, and such other knowledge as he may be willing to communicate concerning the best mode of constructing and securing doors, to be employed for cells or chambers, intended for the reception of patients in a state of mental derangement."[46] Standard locks, apparently, failed to address the committee's concerns about security. They wanted something sturdy enough to secure the doors of insane patients. While "massy bolts" commonly used in asylums locked patients in their rooms, both their appearance and their "harsh ungrateful sound" emphasized confinement over care. The committee considered Thornton the most likely source of knowledge about any improvements in lock technologies that would be appropriate to confine insane patients. While the Committee on Fastenings waited to hear back, they assigned to Thomas Wistar, Ellis Yarnall, and Joseph M. Paul the tasks of investigating whether or not an existing lock design, the Lancaster lock, could be modified to meet their needs. In the process, they also had to determine what sort of lock "may be proper to be adapted" to secure the bedroom doors.[47]

Then, at the end of the month, the Committee on Fastenings reported back to the building committee that Thornton had recommended "the Pendulum lock invented and sold by Samuel Goodwin of Baltimore." Goodwin's pendulum lock was a recent invention that had promised improved security. In 1806, Goodwin and Richard Gaines had applied for and received a patent for a pendulum lock, a type of mortise lock that included a spring or catch that ensured the locking mechanism could not be turned when locked. It also included a protective plate and a keyhole for added security.[48] Three years later, Goodwin received a patent for an improved pendulum lock.⁴⁹ Acting on Thornton's advice, the building committee instructed Wistar, Yarnall, and Paul to procure a pendulum lock and to evaluate it along with the Lancaster lock to determine if either was suitable. Along with their report, they were to submit samples of both the Lancaster and pendulum locks for the building committee's inspection.⁵⁰

Whatever the improvements in Goodwin's pendulum lock, when the Committee on Fastenings was finally able to inspect Goodwin's device, they decided it did not meet their requirements.⁵¹ Neither, however, did the Lancaster lock. Instead, they decided to design their own lock. In August, they submitted the pattern for the lock. The building committee reviewed the pattern and approved it, and then instructed the Committee on Fastenings to find a locksmith who could make forty-seven of them—twenty small locks and twenty-seven large locks—for the southeast wing and the third floor, as well as six keys.⁵²

Just as the SALA had evaluated local blacksmiths for the production of the iron sashes, the Committee on Fastenings now contacted local locksmiths to find one qualified to produce the locks they had designed, in the quantity required and on schedule. The following month, Yarnall reported on behalf of the Committee on Fastenings that they had entered into a contract with John Kennedy to make the forty-seven locks at a cost of two dollars and fifty cents each. Kennedy had agreed to deliver the locks by November 1.⁵³ The committee continued to exercise control over the production and evaluation of the locks. When Kennedy delivered the locks on October 24, the committee tested them to ensure that they functioned properly and as designed. They found three locks and two keys that did not work correctly. Although they paid Kennedy in full for the order, they required him to take the defective locks and keys back and repair them "so as to work well, according to agreement."⁵⁴

Once installed, the building committee must have been satisfied that their newly designed locks met the asylum's requirements for both aesthetics and security. In June 1816, when they ordered the twenty large and twenty small locks for the northwest wing of the asylum, the building committee instructed the Committee on Fastenings that these locks should be "of the same construction as those heretofore procured for the SE wing."[55] The Committee on Fastenings turned again to Kennedy, whose workmanship had met the committee's standards. Two months later, the Committee on Fastenings reported that Kennedy would manufacture the remaining locks. He delivered them to the asylum by November. Once again, when the committee paid Kennedy, they required him to guarantee his workmanship: "It being now agreed, on my part, to repair any of the locks that I have made without any charge—provided they shall have become deficient in consequence of not being made in a workman like manner."[56] The Quakers' ethos of care directed itself to patients unable to care for themselves, and to the materials and artifacts enjoined in that caretaking.

CONCLUSION

What do we gain by lending such equivalence to the medical and material problem choices or solutions taken up by the Quaker asylum builders? And, what do we gain by seeking a symmetry in the roles played by people and materials enlisted in that program of caregiving? For one thing, an alternative explanation for the nature of human-built structures readily emerges as a stylistic choice reattached to actors' ethical and practical priorities, enriching the history of architecture. But that historiographic corrective prompts a wider reassessment of the nature of Quaker caretaking by revealing at least two things. First, we can see the nature of what constituted early nineteenth-century Quaker care of the insane more clearly: providing access to light and air that rendered selectively permeable the boundary between the asylum's inside and outside; providing a sensation of domestic security rather than punitive confinement; and maintaining a sense that however debilitated they might be, the mad still manifest some degree of meaningful volition. Each articulated priority embodies the Quakers' sense of real and possible human engagements with the world, of potential pleasure and contentment; these potentialities are distributed across individuals of differing mental health, perhaps, but entirely foreclosed for no individual.

Second, if we see the solicitous glazed window, the iron sash made to look like wood, or the soft-spoken lock as active participants in that program of caregiving, we see further features of the Quaker ethos that might not otherwise be visible. Before Tuke's retreat and the Philadelphia design inspired by it, asylum walls, windows, and locks actively made their inmates ill, inducing unwanted behaviors. With reforms, materials and structures helped to cure. What building features were worthy of innovation to the Friends' Asylum committee members, and what was then invented or repurposed in the world of building materials, literally played a role in making the experience of madness a Quaker one, and thus a relatively positive one. For example, the iron sash confined but did not recriminate; the quiet lock secured but did not redundantly remind the inmate of his or her incarceration. Perhaps Quaker-asylum madness was in fact a positive one in some absolute sense as well, as it made not just the Subcommittee on Light and Air, the Committee on Fasteners, and the materials they handled more caring, but it made of the inmates' insanity—lovingly perceived, experientially designed—itself an identifiably Quaker experience.

NOTES

1. Minutes of the building committee, December 25, 1813.
2. In the early nineteenth century, *insanity* existed along a host of other terms, including *madness, lunacy, mania,* and *hysteria*. We find insanity along with a host of other terms in both the medical literature and the managerial materials. For example, Isaac Bonsall, the first superintendent of the Friends'

Asylum, described patients as insane and suffering from insanity, and we find additional terms such as *spasms, very noisy, crazy, depression of mind*, and *derangement*, among others. See Isaac Bonsall, "Superintendent's Daybook," vol. 1, Friends Hospital records (MC 1261), Quaker & Special Collections, Haverford College, Haverford, PA, 1817–1820. For a broader survey of the term *insanity* in its contemporary usage, see Andrew Scull, *Madness in Civilization: From the Bible to Freud, from the Manhouse to Monder Medicine* (Princeton, NJ: Princeton University Press, 2015), especially chapters 3–7.

3. William Tuke in York along with Philippe Pinel in Paris are often credited with having created this new moral treatment and having popularized it both among asylum keepers and designers and among the public. See Anne Digby, "Changes in the Asylum: The Case of York, 1777–1815," *Economic History Review* 36 (1983); Anne Digby, "Moral Treatment at the Retreat, 1796–1846," in *The Anatomy of Madness: Essays in the History of Psychiatry*, eds. W. F. Bynum, Roy Porter, and Michael Shepherd (London: Tavistock Publications, 1985), 3–51.

4. Anonymous, *Account of the Rise and Progress of the Asylum, Proposed to be Established Near Philadelphia, for the Relief of Persons Deprived of the Use of Their Reason With an Abridged Account of the Retreat, a Similar Institution Near York, in England* (Philadelphia, PA: Kimber & Conrad, 1814), 74. The Quakers referred to the community of people who would live at the asylum, whether receiving treatment or administering care or managing the asylum, as "the family." The first superintendent regularly referred to the patients, the workers, and even his own family who lived at the asylum as "the family."

5. Drawing on Latour's actor-network theory, Albena Yaneva has argued forcefully for the agency of things in the context of design and building. See Albena Yaneva, "Making the Social Hold: Towards an Actor-Newtork Theory of Design," *Design and Culture* 1 (2009): 273–88

6. The chapter by Scott Knowles and Jose Torero in this volume traces another fascinating instance of an old material being made new when invested with a new set of values and ethics. And how those values are contested and thus the very nature of wood, what it is, depends on the ethics brought to bear on it. See Scott Gabriel Knowles and Jose Torero, "Plyscrapers, Gluescrapers, and Mother Nature's Fingerprints" in this volume.

7. Barry Allen has argued that because the solution to any technological problem is not technologically determined, the resulting technological artifact is the result of multiple aesthetic choices and perceptual forces. In this way, the final look, feel, and sound of any artifact is integral to the design of that artifact and as important as the technical or engineering problems. On this

analysis, form does not follow function but rather plays an important role in shaping function. See Barry Allen, *Artifice and Design: Art and Technology in Human Experience* (Ithaca, NY: Cornell University Press, 2008), especially chapter 4.

8. Anonymous, "Lunatic Asylums," *Edinburgh Review or Critical Journal* 28 (1817): 453.

9. This chapter departs to a significant degree from historical archeology by considering the inescapable coproduction of materials (in this case, as caring objects) and the identities of their users (as caring subjects); see James Deetz, *In Small Things Forgotten: The Archeology of Early American Life* (New York: Anchor Press, 1977).

10. Allen makes an analogous argument in analyzing bridges into New York. He shows how these large-scale civic projects cannot be bifurcated into engineering solutions to technical problems and subsequent aesthetic choices. Because aesthetic choices shape technological problems, the two are necessarily and inextricably intermingled. See Allen, *Artifice and Design*, chapter 4.

11. This chapter, like any study on asylum architecture and planning, owes a debt to Michel Foucault's early work. See Michel Foucault, *Madness and Civilization: A History of Insanity in the Age of Reason*, trans. Richard Howard (New York: Vintage Books, 1988). See also the more recent, full translation of the original, Michel Foucault, *History of Madness*, trans. Jonathan Murphy and Jean Khalfa (New York: Routledge, 2006). More recent scholarship on asylum design and architecture has challenged and refined Foucault's conclusions. See, for example, Lindsay Prior, "The Architecture of the Hospital: A Study of Spatial Organization and Medical Knowledge," *British Journal of Sociology* 39 (1988): 86–113; Christine Stevenson, *Medicine and Magnificence: British Hospital and Asylum Architecture, 1660–1815* (New Haven, CT: Yale University Press, 2000); Carla Yanni, *The Architecture of Madness: Insane Asylums in the United States* (Minneapolis: University of Minnesota Press, 2007); and, more generally, Andrew Scull, *Madness in Civilization*. Unsurprisingly, the York retreat and English asylum have received considerable attention. Representative of this literature is work by Anne Digby and by Barry Edginton: Digby, "Changes in the Asylum"; Digby, "Moral Treatment at the Retreat"; Barry Edginton, "The Well-Ordered Body: The Quest for Sanity Through Nineteenth-Century Asylum Architecture," *Canadian Bulletin of Medical History* 11 (1994): 376–86; Barry Edginton, "The Design of Moral Architecture At the York Retreat," *Journal of Design History* 16 (2003): 103–17; and Barry Edginton, "A Space for Moral Management: The York Retreat's Influence on Asylum Design," in *Madness, Architecture, and the Built Environment. Psychiatric*

Spaces in Historical Context, eds. Leslie Topp, James E. Moran, and Jonathan Andrews (New York: Routledge, 2007), 85-104.

12. Annmarie Adams makes related observations in the introduction to her *Medicine by Design*, where she worries that histories of hospital design tend to see the development of interior spaces as the logical and conceptual result of the rise of medical theories; e.g., Koch's germ theory. In that model, architecture becomes a passive expression of therapeutics. While she stops short of considering hospital design and the materials that realize that design as active participants in therapeutics, she does grant design a productive role in therapeutics. See Annmarie Adams, *Medicine by Design* (Minneapolis: University of Minnesota Press, 2008), Introduction.
13. Samuel Tuke, *Description of the Retreat, an Institution Near York, for Insane Persons of the Society of Friends* (Philadelphia, PA: Isaac Pierce, 1813), 70.
14. Tuke, *Description of the Retreat*, 66.
15. Anonymous, "Lunatic Asylums," 455.
16. Tuke, *Description of the Retreat*, 35.
17. Anonymous, *Account of the Rise and Progress*, 48.
18. Anon., *Account of the Rise*, 41-42.
19. William Stark, *Remarks on the Construction of Public Hospitals for the Cure of Mental Derangement* (Edinburgh: James Ballantyne, 1807), 7-8.
20. Stark, *Remarks*, 1.
21. Thomas Scattergood, "Diaries," vol. 13, Thomas Scattergood Diary, Scattergood family papers (MC 1100), Quaker & Special Collections, Haverford College, Haverford, PA.
22. Anon., *Account of the Rise*, 32.
23. Tuke, *Description of the Retreat*, 68. When the Philadelphia Quakers printed an abridged account of Tuke's *Description*, they included this section on windows and cold; Anon, *Account of the Rise and Progress*, 34.
24. Samuel Tuke, *Practical Hints on the Construction and Economy of Pauper Lunatic Asylums* (London: William Alexander, 1815), 39.
25. Anon., *Account of the Rise and Progress*, 34.
26. The author of the review article in the *Edinburgh Review* was particularly concerned that patients' bedroom windows were typically without glazing. See Anon., "Lunatic Asylums," 440, 442.
27. Samuel Tuke emphasized the modest appearance of the retreat even while quoting Delarive's description of the retreat as a "large rural farm." Tuke, *Description of the Retreat*, 62.
28. Anon., "Lunatic Asylums," 462.
29. Tuke, *Practical Hints*, 35-36.

30. Tuke, 37.
31. Tuke, 36–37. In other places, Tuke describes the panes as six inches by seven and a half. See Tuke, *Description of the Retreat*, 64.
32. Tuke, *Description of the Retreat*, 65.
33. Minutes of the building committee, Friends Hospital records (MC 1261), Series 1: Administrative Records, Subseries 9, Quaker & Special Collections, Haverford College, Haverford, PA, October 11, 1813.
34. Minutes of the building committee, Friends Hospital records (MC 1261), Series 1: Administrative Records, Subseries 9, Quaker & Special Collections, Haverford College, Haverford, PA, December 25, 1813.
35. Minutes of the building committee, Friends Hospital records (MC 1261), Series 1: Administrative Records, Subseries 9, Quaker & Special Collections, Haverford College, Haverford, PA, February 12, 1814.
36. Cecil D. Elliott, *Technics and Architecture* (Cambridge, MA: MIT Press, 1993), 126–27.
37. Board of managers minutes, March 28, 1814.
38. Anon., *Account of the Rise and Progress*, 10.
39. Minutes of the building committee, Friends Hospital records (MC 1261), Series 1: Administrative Records, Subseries 9, Quaker & Special Collections, Haverford College, Haverford, PA, March 12, 1814.
40. Minutes of the building committee, Friends Hospital records (MC 1261), Series 1: Administrative Records, Subseries 9, Quaker & Special Collections, Haverford College, Haverford, PA, June 4, 1814.
41. Minutes of the building committee, Friends Hospital records (MC 1261), Series 1: Administrative Records, Subseries 9, Quaker & Special Collections, Haverford College, Haverford, PA, September 17, 1814.
42. Receipt dated July 15, 1815. Friends Hospital records (MC 1261), Series 3: Financial Records, Subseries 1, Quaker & Special Collections, Haverford College, Haverford, PA.
43. At the meetings on September 2, 1815, and October 14, 1815, the building committee confirmed that the garret and transom windows should use moveable iron sashes rather than iron bars. At the meeting on August 20, 1814, the building committee had recommended the use of iron bars for both.
44. Two surviving receipts indicate that the SALA paid Benjamin Jones $1297.76 for iron sashes. While these receipts account for the majority of the windows (apparently all the exterior windows), they do not include all the windows in the asylum. Receipts for many of the smaller, probably transom, windows are missing. See receipts dated August 15, 1816, and October 20, 1815. Both in Friends Hospital records (MC 1261), Series 3: Financial Records, Subseries 1, Quaker & Special Collections, Haverford College, Haverford, PA.

45. Tuke, *Practical Hints*, 35.
46. Minutes of the building committee, Friends Hospital records (MC 1261), Series 1: Administrative Records, Subseries 9, Quaker & Special Collections, Haverford College, Haverford, PA, April 1, 1815.
47. Minutes of the building committee, Friends Hospital records (MC 1261), Series 1: Administrative Records, Subseries 9, Quaker & Special Collections, Haverford College, Haverford, PA, April 8, 1815.
48. I thank Robin Kaller of Kaller Historical Documents, Inc., for providing me with a photograph of the letter describing Goodwin and Gaines' pendulum lock.
49. A fire at the US Patent Office destroyed the documents that described this second lock. Consequently, I do not know what improvements Goodwin introduced in this second invention. I know only that on July 7, 1809, Goodwin received a patent for a new or modified pendulum lock.
50. Minutes of the building committee, Friends Hospital records (MC 1261), Series 1: Administrative Records, Subseries 9, Quaker & Special Collections, Haverford College, Haverford, PA, April 28, 1815.
51. For two months, the Committee on Fastenings reported that it had not yet received Goodwin's pendulum lock.
52. Minutes of the building committee, Friends Hospital records (MC 1261), Series 1: Administrative Records, Subseries 9, Quaker & Special Collections, Haverford College, Haverford, PA, August 12, 1815.
53. Minutes of the building committee, Friends Hospital records (MC 1261), Series 1: Administrative Records, Subseries 9, Quaker & Special Collections, Haverford College, Haverford, PA, September 2, 1815.
54. Receipt dated October 24, 1815. Kennedy's note and agreement to repair the locks is on the back of the receipt and dated October 30. Friends Hospital records (MC 1261), Series 3: Financial Records, Subseries 1, Quaker & Special Collections, Haverford College, Haverford, PA.
55. Minutes of the building committee, Friends Hospital records (MC 1261), Series 1: Administrative Records, Subseries 9, Quaker & Special Collections, Haverford College, Haverford, PA, June 22, 1816.
56. Receipt from November 16, 1816. Friends Hospital records (MC 1261), Series 3: Financial Records, Subseries 1, Quaker & Special Collections, Haverford College, Haverford, PA.

CHAPTER SEVEN

CULTURAL FRAMES

Carbon-Fiber-Reinforced Polymers, Taiwanese Manufacturing, and National Identity in the Cycling Industry

Patryk Wasiak

INTRODUCTION

It is commonplace today to speak of a globalized industrial market where production and consumption, design and marketing, flow as far and as fast as international trade law will allow. But a critical history of new industrial materials, following their development, deployment, and international reputations, introduces a set of complex cultural factors into this discourse of a unified commercial world. This chapter considers how the reputation of a modern industrial product may not only be inseparable from variable audience demands for functionality, but from audience ideas of national technological prowess.

Vitally, those two kinds of expectations may well be inseparable from one another, with conceptions of product value deriving from ethnic stereotypes both negative and positive. That is to say, in product designs and material choices, producers are seen by different markets as carrying forward both favorable and unfavorable national habits. This chapter takes, as its case, the twenty-first century introduction of Taiwanese carbon-fiber-reinforced polymer (CFRP) components into the production of racing bicycles, a commercial expedient by European bicycle makers that has introduced so-called Asian manufacturing to a product familiar to many buyers as "purely" Italian or French artisanal output. Where consumers have detected ingenuity and workmanship in the resulting bicycles, and where they have felt a threat from perceived weaknesses in Asian industrial practices, has much to tell us about the layered nature of national identity, technological confidence, and global flows of skills and goods today.

This case can also tell us about the very nature of foreignness in a world of industrial exchange: What exactly makes a place or people seem "other" today? Taiwan is, by any measure, a global potentate in the provision of CFRP bicycle components, achieving an unprecedented scale for this very advanced manufacturing undertaking. Yet this technical success did not bring about seamless uptake of Taiwanese components by global markets. Nor are end users of high-end products the only actors we need to follow to understand this landscape. That European bicycle makers, turning from conventional metal-alloy elements to the new material, saw their task as reconciling international production operations and core brand values is itself telling about the cultural complexity of multination manufacturing. That it was athletic prowess and not some other kind of functionality being sought in these bicycles introduces an even richer set of questions to this inquiry: the performance of the new components could empower or constrain, quite literally, the buyer of the globalized bikes. How a new industrial material comes to seem reliable in the twenty-first century (or fails to achieve that reputation), how its makers come to seem competent and trustworthy (or not), and how either comes to seem familiar or foreign are the subjects of this chapter.

First, it is necessary to briefly introduce CFRP. This material, often referred to simply as carbon fiber, includes intertwined carbon filaments bonded with heat-cured resin into a structure with a strength-to-weight ratio factor much better than that of metal alloys.[1] This material was developed by British aerospace industry engineers in the 1960s and was ultimately disseminated in industrial applications and high-end sporting-goods manufacturing. In the late twentieth century, CFRP gained favorable publicity as a new "wonder material," which promised advancement for aerospace and automotive industries, military technologies, and sporting goods. Its main advantages were seen to consist of better strength and a stiffness achieved through construction lighter than the same design made with metal. The public recognition of CFRP as a material of the future with significant advantages over metal, especially its promise to build lighter and faster vehicles, shows similarities with the public life of aluminum in the interwar period.[2] Yet, there is a significant difference between the histories of the two materials. In the course of about three decades, aluminum became very cheap. It moved into use in the manufacture of mundane objects such as soda cans, while CFPR, after more than four decades, is still very expensive to deploy in manufacturing. It is used primarily in applications where cost is not an overriding concern, such as in military aircraft, racing cars, and premium sporting goods.

Used as a material to manufacture racing bicycles, CFRP offers a range of advantages over metal alloy.[3] A bicycle frameset, wheels, and other components such as a handlebar or crank set made with CFRP offer better stiffness and lighter weight that contribute to an athlete's performance. Stiffness of bicycle components causes any power input by the cyclist into pedals to be better transmitted into speed, since less power is lost through the small instances of bending to which less stiff materials are prone. The second advantage is lightness, which significantly decreases the power required to accelerate the bicycle, especially on ascending terrain. The third benefit is that carbon fibers can be easily molded into any desired shape. Thus, a frameset manufactured with CFRP could be designed, taking aerodynamic efficiency into account, while the design of a metal-alloy frameset would be limited by the shape of available tubing. To summarize: CFRP significantly improves athletes' performance in cycling by offering better power transmission and better climbing, and by facilitating the building of aerodynamic components. Aleksander J. Subic and Steve J. Haake, in their book on the role of technology in sport, point out that the introduction of CFRP as a material in racing-bicycle frames had one of the most spectacular impacts of engineering on sport in recent years.[4] Yet, this article shows that the introduction of CFRP cannot be accounted for simply as a high-tech material replacing an obsolete one, but rather offers insight into cultural values embedded in different materials and tensions between cultural identity and commercial expedience.

I discuss the introduction of CFRP in the cycling industry in light of the values related to the ingenuity of design, to the mastering of building artifacts with metal alloy, and to strong interdependence of the country of origin and product attributes—all as shaped during the late nineteenth century, a peak period of industrial development, primarily in France and Italy. As I will further show, those values persisted for most of the twentieth century. Until the 1990s, the cycling industry was dominated by French and Italian manufacturers who, through many generations, built their brand identities by emphasizing the values of ingenious design, artisanal manufacturing of metal alloy, and style. From the early twentieth century, tradition became a core brand value in the industry. Moreover, not only the bicycle's manufacturer but also the country of origin of the metal alloy tubing used to build the bicycle brought significant added value, in the eyes of many.

The popularity of bicycles labelled "Made in Italy" and "Made in France" was a result of the notion of the high quality of these particular technological artifacts and the perceived attention to the design of consumer goods generally by manufacturers from both countries. While discussing this process of reputation-building, I focus on Italian manufacturing companies that developed the most sophisticated brand identities as manufacturers of bicycles designed with long experience, aesthetics, and a strict reliance on Italian-made tubing. With the emergence of contemporary global supply-chain routes and the popular practice of using offshore, original equipment manufacturers (or OEM), also growing at the end of the twentieth century, such traditional values developed by European companies became blurred: high-end bicycles offered for sale by European companies are today manufactured by Taiwanese subcontractors. Both reputational issues discussed in my article—the stereotypes of claimed high quality of made-in-Italy and low quality of made-in-Taiwan products, and the CFRP bicycle frameset as a Taiwanese success story—show that the manufacturing of technological artifacts is not easily characterized as expressing any sort of universal knowledge.[5] (See also Tiago Saraiva's essay in this volume.)

The case of high-end bicycle production also helps us track the role of the nation as a unit of analysis among those who gauge the cultural significance of technological developments. This is, of course, not new to recent eras. Historians have mapped the coemergence of national identities and the so-named Industrial Revolution in nineteenth-century Europe.[6] To quote what is likely Benedict Anderson's best known concept in this arena, nations were built as imagined communities based on several factors such as common language and culture, but also on imagined scientific and technological prowess.[7] Advancements in technologies and infrastructures such as steamships, railways, and metal constructions of unprecedented scale (for instance, the Eiffel Tower) came to be recognized as a result of the originating nation's capacities and were included in national narratives. The most spectacular contest between different nations' technological prowess could be observed at international exhibitions, where national pavilions functioned as lavish shrines to technological progress. Such narratives of technological achievements always require a negative point of reference; that is, the projection of other nations with supposedly less prowess in science and technology. For example, while discussing an instance of German motorways (Autobahn) during both Nazi and postwar periods, Thomas Zeller shows how transport infrastructure could be deeply engaged in a supremacist national narrative.[8] Similarly, Saraiva shows how animals, literal "fascist pigs," also became technological objects nationalized by the Italian Fascist regime.[9] During the Cold War period, the Soviet Union was popularly recognized in Europe and the United States as a state notorious for mindless copying of Western technologies. This notion was fueled by traditional stereotypes concerning the civilizational backwardness of Russian folk, and helps us see that an absence of ingenuity could also be actively constructed by those seeking to degrade another culture. When Asian technological products—primarily cars and consumer electronics—entered markets abroad in the 1960s and 1970s, Western consumers recognized them as poorly made and imitative products that could succeed only because of their cheapness. Such assumptions were fueled by preexisting stereotypes concerning the lack of any historical scientific and technological practices in Asia.

The entrance of Taiwanese CFRP bicycles into global markets late in the twentieth century provides another case of intersecting audience ideas about producer nations' scientific and technological ingenuity and prowess (or the lack of thereof) and the categories of nationhood and race. Here, I refer to such judgments as *country-of-origin stereotypes*—a term extensively used in consumer-culture studies to understand consumer perception of the country where a commodity was designed and manufactured as an important cue in the estimation of product quality.[10] Country-of-origin (henceforth

COO) stereotyping is an instance of knowledge related to consumption that includes technical, mythological, and evaluative components, which we may see as simultaneously producing both a judgment about those factors and a sense of the nationality of self and others.[11] As a number of studies of the impact of COO on consumers' decisions show, the consumer's perception of material quality and the concept of national identity strongly correlate.[12]

To explore these ideas, this chapter is structured as follows: In the first part, I reconstruct the origins of traditional cultural values of the bicycle industry based on long experience in metal-alloy tubing manufacturing and conceptions of Italian and French nationhood; I next outline the role of the Taiwanese government and bike manufacturers in the introduction of CFRP as a Taiwanese flagship high-technology product; and in the final section, I reconstruct the tensions related to the offshoring by European bicycle firms to Taiwan, and how COO has recently functioned as an element of consumers' evaluation of CFRP bicycles. Working with bicycle manufacturers' catalogs, websites, and trade journals, and studying popular blogs and internet forum threads, we see that technological prowess is rarely invoked without some reference to nationality. Global flows of goods and labor aside, or perhaps exactly because of those flows, a bicycle component's country of origin plays a substantial role in product evaluation by consumers.

THE EUROPEAN INDUSTRIAL AGE AND THE METAL-ALLOY BICYCLE

In this section, I discuss the emergence of cycling as a sport in the context of nation-building in Europe and the importance of manufacturing bicycles with metal alloy, a technology eventually replaced by CFRP. The general idea of competitive cycling and the design of racing bicycles aimed at providing athletes with the best possible speed performance were developed in western Europe in the first decade of the twentieth century.[13] The concept of bicycle races based on competition among trained athletes, as well as among the bicycle manufacturers who provided the athletes with high-performing machines, was strongly influenced by the discourse of industrial modernity. The emergence of sport cycling, manifested through races in Europe, was embedded in the broader cultural context of the era. First, several high-profile stage races such as the Tour de France and Giro d'Italia were run through multiple picturesque areas of the given country. They encouraged the discovery by new audiences of the attractive countryside and promoted tourism, a new form of leisure available for the masses in this era. Moreover, races held in difficult terrain, especially passing through mountain roads, were used to display both athletes' abilities and the quality and durability of bicycles used by such superlative consumers. Currently, we rather heavily identify the engine-powered car as a symbol of the industrial development of the era.[14] However, the historical popularity of bicycles as a symbol of achievements of industrializing countries should not be underestimated. A racing bicycle was a human-powered technology in which success (i.e., the winning of the race) depended simultaneously on the design and quality of the technological artifact in use and athletes' strength and stamina.[15]

The growth of national ideologies of France and Italy, two countries where competitive cycling was pioneered, embody this sense of national commitment to developing identity and skills simultaneously. Hugh Dauncey and Geoff Hare, in their book on the history of the Tour de France (an event established in 1903), reconstruct this context: "Cycling and the Tour were instruments for the definition of France and for the improvement of French society through technology (the industrially mass-produced cycling machine) and the athletic prowess of her menfolk."[16] The Tour de France was a manifestation of French technology and culture as inseparable undertakings. Until the 1950s, when the Tour became an international sporting event, French cyclists were racing with French manufactured bicycles on French roads. Roland Barthes, in his famous essay "The Tour de France as Epic," identified the Tour as a substantial part of French contemporary mythology.[17] Another aim of such races was the exploration of the nation-state's provinces as places that could be further explored by dwellers of main cities who, at the same time, learned about tourism as a new form of leisure. For this period, the basic rule of race organization for national tours was that the route encircled the state territory; thus, central and local authorities were able to use races to strengthen brand identities of both states and local regions as attractive tourist destinations. In Italy, which became a nation-state in 1861, the Tour of Italy (Giro d'Italia) established in 1909 played a significant role as a nation-building event.[18] The engagement with cycling as a credible sport and the industry's value in political discourse are well illustrated with the name of one of the oldest Italian bicycle manufacturers: Wilier (established in 1906). The company name is an acronym that stands for "W (Viva) l'Italia liberata e redenta" (Long live Italy, liberated and redeemed).

Manufacturers' contributions to national identity and solidity relied on more than patriotic fervor, of course: the performance of bicycles used in such races was dependent on the technology of metal alloy, and bicycle manufacturers of the day built their brand values on mastering metallurgy. The industrial development of the bicycle as the technological artifact we know today has been discussed in depth by Wiebe E. Bijker.[19] The bicycle construction, which includes a diamond-shaped frame and two equal-sized wheels, was stabilized at the end of the nineteenth century. The construction of the frameset of such bikes required both good-quality metal-alloy tubing and sophisticated alloy-manufacturing techniques. Both features, the quality of the raw material and manufacturing skills, became, to refer to Arjun Appadurai, substantial elements of knowledge that went into both the production and consumption of bicycles for several decades.[20] As Bijker points out, the first entrepreneurs of the cycling industry were blacksmiths and mechanics.[21] Further, small workshops of those entrepreneurs who specialized in the metallurgy of bicycle manufacture were the origin of subsequent larger industrial companies. Several of those companies survive to this day.

The basic component for bicycle frames in the early era of mass production was steel alloy tubing. In *Bicycling Science*, David Wilson claims that the cycling industry pioneered the development of thin and strong metal-alloy tubing, and experiences with building bicycle tubing subsequently contributed to other metal-alloy industries.[22] Bruce D. Epperson, while discussing the cycling industry in the US, describes the pursuit of steel tubing suitable for the diamond-frame bicycles: "The thin-walled, high-strength tubes now needed were more like shotgun barrels than the relatively thick and heavy backbones of the Ordinary [the bicycle with a large front wheel]."[23] He also claims that the lack of suitable tubing, known at the time as the "tube famine," was a significant constraint in the manufacturing of bicycles in the 1890s.[24] Soon thereafter, in the European cycling industry, the supply of metal-alloy tubing was dominated by two suppliers: the English Reynolds Cycle Technology (established in 1898), which was later joined by the Italian Columbus (1919). In 1934, Reynolds patented its famous Reynolds 531 steel-alloy tubing. It was named after the ratio of the other materials used in the steel alloy: manganese, molybdenum, and silica. The current advertising slogan of Columbus, "The soul of cycling since 1919," refers to the fact that, along with Reynolds, its tubing became the industry standard for high-end racing bicycles for several decades. Browsing European bike manufacturers' vintage catalogs shows that in the cases of virtually all high- and mid-range bicycles, information on tubing manufacturers, mostly with specific tubing model (and sometimes accompanied by manufacturer's logo), is included in technical specifications.[25] The use of technologically advanced materials supplied by renowned contractors was an important element of bicycle firms' marketing strategies. Analysis of such promotional materials shows that manufacturers presented a high-end bicycle to consumers as a technological artifact built upon well-chosen materials, reliable components, and the selling firms' expert knowledge about both.

Several founders of bicycle manufacturing companies had a personal background in metal-processing industries, confirming our understanding of specialized material knowledge as integral to manufacturing success. For instance, Jean Pequignot Peugeot, a founder of the Peugeot company (established in 1858), was the owner of steelworks who expanded his product range with penny-farthing bikes (those with a huge front wheel) known as Cycles Peugeot from 1882 onward, as well as cars (Peugeot's cycle and car manufacturing branches separated in 1926). In Italy, Tullio Campagnolo, founder of the respected Campagnolo company (established in 1933), was the son of an ironmonger who began his experiments in his father's hardware store.[26] Later, he became perhaps the most widely known bicycle-component designer of the twentieth century; among other things, he invented several derailleur systems.[27] Further, several other founders of bicycle manufacturers had personal experience with professional cycling teams as riders and mechanics, thus gaining knowledge on the practice of using and servicing bicycles for competition. Ernesto Colnago, the founder of the Colnago company (established in 1954), had both experience as a rider and a bike mechanic in several Italian cycling teams. Those two elements—experience in metal industries and personal knowledge regarding cycling—were seen as essential to effectual design and treated as the core values of particular bicycle manufacturers' brands.

Until the 1980s, credible racing-bicycle manufacturing remained primarily an artisanal production along these lines, emphasizing knowledge closely held in the hands of those responsible for design and manufacturing, known as *master frame builders*.[28] Bikes were assembled in workshops and advertised with omnipresent assurances that each was hand built in France or Italy. In the cycling industry, bicycles manufactured in both countries became recognized as premium sporting goods superior to bicycles made in other regions. Here, the value of French and Italian artisanal production also drew on a broader image of both countries as producers of high-quality, well-designed luxury goods. As James B. Twitchell writes in his book on the recent social construction of luxury, "If you are going to flaunt something . . . chances are you will be doing it with something French or something Italian."[29] Twitchell analyzes the contemporary world of luxury commodities, but the notions of "made in France" and "made in Italy" as high-quality and fashionable products could be tracked to the early twentieth century and the development of international mass commerce.[30] Italy especially became recognized in this period among Western audiences as a country of origin of consumer goods, routinely manufactured taking account of aesthetics.[31] Colnago's slogan used on their website, "Colnago's manufacturing philosophy combines innovation, design and beauty," could be easily applied by Italian manufacturers of sports cars or home appliances. US companies such as Trek, Cannondale, and Specialized, who entered the cycling market in the 1970s and 1980s, applied similar marketing strategies, and emphasized the role of handicraft with the slogan "Handmade in the USA," emphasizing the quality of US artisanship, and referencing the same values supposedly driving European manufacturers. For the sake of brevity, I omit a complex history of US firms entering the world markets, not only with "Handmade in the USA" claims but with the innovation of the mountain bicycle, a product bearing strong references to US-style athleticized leisure embodied in skateboards, surfboards, and mountain bicycles.[32] But we can certainly acknowledge the general aim among sporting-goods makers to associate their technological and design priorities with their nationalities.

Aforementioned notions of the quality of artisanal manufacturing prevailed for several decades in the industry, where the basic material for virtually all bicycle components was metal alloy. The only massively adopted material innovation involving metals before the introduction of CFRP was the introduction of aluminum, applied as a material for bike components other than framesets from the 1920s and 1930s onward by European manufactures in high-end bicycles. In *Aluminum Dreams*, Mimi Sheller refers to the meaning of aluminum as a "speed metal" and discusses its applications, mostly in transportation where its lightness contributed to the increase of speed.[33] Aluminum, with its lower weight than steel, could significantly improve the cyclist's performance, especially while on mountain ascents. For instance, the French company Mavic, now dominant in bicycle wheels, developed wheel rims made with duralumin (an alloy of aluminum and copper), named Mavic Dura, which significantly dropped the weight of bicycles used in the Tour de France in 1931.[34] However, framesets made with aluminum only came to the fore in the 1980s. This decade was also a period of several experiments with other materials. Aside from aluminum, titanium was used as a material for several high-end frames, but the high price of this metal and very difficult processing requirements meant that it was not widely applied in the industry. Some companies experimented with other unconventional materials such as fiberglass, exotic woods, and several plastics as components for framesets, with little success. Among these and other attempted innovations in materials applied for bicycle components, CFRP gained popularity as the most promising. It was understood in an immediate sense to carry the most advantages for athletes' performance. However, the recognition by racing-bike makers and buyers of such benefits alone does not explain how CFRP came to be seen as superior to other options. As I will show in the next section, the application of CFRP was also a result of active policy-making by a range of social actors, mostly Taiwanese government figures and manufacturers who sought to join the global world of advanced technologies with CFRP manufacturing as their flagship product.[35]

THE INTRODUCTION OF CFRP

The first carbon-fiber-reinforced polymer was developed by engineers at British Aerospace (a major British defense contractor) in the 1960s, as a lightweight material that could be applied to improve the design of military and civilian aircraft and space vehicles.[36] The main advantage of this composite material in that context was a strength-to-weight ratio significantly better than that of steel or aluminum. The main limitation of the CFRP was its very expensive manufacturing process and relatively low impact resistance. In the 1970s, CFRP was applied on a small scale in the aerospace industry. In the next decade, with decreasing manufacturing costs, it started to be widely applied in civil engineering, as well as in sports car design, where cost was not a primary concern. In popular science discourse, CFRP became recognized as the next "wonder material," which promised "superstrong, lightweight substitutes for present components."[37] Similar to aluminum in the 1920s, CFRP promised high-tech machines that would achieve better speed by surpassing limitations caused by the weight of metal. There is an important difference between CFRP and aluminum. The cost of production of the latter in a commercial capacity quickly declined because of the development of a massive system of extracting and processing aluminum ore by the Alcoa (Aluminum Company of America) Corporation, as that massive firm pursued economies of scale. Similarly, another wonder material—plastic—found favor in manufacturing sectors comfortable with investment and growth. Quickly manufactured in abundance at the time, it rapidly became a part of everyday life in many industrialized cultures.[38]

Contrary to those materials, more than four decades since its first appearance, production of CFRP is today still very expensive, and its application in consumer technologies is limited to premium goods. In 2008, about 46 percent of the world's CFRP was applied for industrial purposes in civil engineering, electronics defense, and energy sectors; about 28 percent in the aerospace industries;

and 26 percent in sporting- and leisure-goods manufacturing.[39] CFRP was introduced in the design of several sporting goods where the low weight and material stiffness had an impact on the athlete's performance and price could be cast as a secondary concern; for instance, in racing bicycles, yachts, rowing boats, golf clubs, and baseball bats.[40] Experiments with the application of CFRP in the bicycle industry started on a very small scale at two European companies. In 1981, Colnago build an experimental CX-1 Pista track race bicycle. A single CX-1 frameset was manufactured by the company in-house with a monocoque piece of molded CFRP. In 1986, the French company Look developed a hand-built KG-86 frame. It was more traditional in design: a diamond-shaped frame in which carbon tubing was jointed with aluminum lugs. CFRP tubing for this design was provided by French aerospace company Toulon Var Technologies.[41] Greg LeMond won the Tour de France in 1986 with the Look KG-86 bicycle. This Tour became famous as the first race won by a US citizen, but it was also the first major cycling race won with the CFRP bicycle. Look, however, didn't market KG-86, since the CFRP frame-manufacturing process was too expensive to be included in any mass-produced bicycle. Only a limited run of this model was distributed to professional teams.

The first mass-produced CFRP frameset was marketed in 1987, however, by the fast growing Taiwanese company Giant, founded in 1972, in a set of circumstances that subverted the idea that high-tech materials and luxury product lines necessarily indicated small-scale production. To understand how a relatively new Taiwanese company managed to diffuse this innovation, we need to go back to the 1950s. In doing so, we can detect a long-standing set of private and public intentions to harness both economic and scientific expertise in the interest of national industrial growth. Currently, Taiwan is known as the world's leading manufacturer of consumer electronics components and a force in the manufacturing of plastics and composites, as well as a major supplier of bicycle components. In the 1950s, Taiwan's rapid industrialization took place, with significant support from the government to lay the foundation for such long-lasting and growing influence across global markets. In 1958, Taiwan's government hired consultants from the Stanford Research Institute to recommend industries that would suit Taiwan's comparative advantages. Consultants advised national leaders to stress plastics, synthetic fibers, and electronic components—all of which soon became staple exports.[42]

The most spectacular Taiwanese success story in the plastics industry became Formosa Plastics (established in 1954), which currently competes in global markets with US potentate DuPont. In the 1970s, well practiced in the scale-up of manufacturing based on new materials, Taiwanese companies started to develop the technology of producing CFRP as well as ready-made CFRP goods. This was, in many regards, a successful effort: Taiwanese companies such as Formosa Plastics and Taiwan Electric Insulator (opened in 1978) today produce a large part of the world's CFRP material output.[43] One of the first marketed CFRP Taiwanese products was a tennis racket; the use of the material in bicycles was soon deemed a reasonable choice. Taiwanese companies had started manufacturing bicycles for export in the 1960s. The main constraint of marketing in this sector was a bad reputation of poorly manufactured "Made in Taiwan" bicycles on foreign markets. In his book on Taiwanese industrial development, Wan-wen Chu claims that to increase the reputation of Taiwanese bicycles worldwide in the 1970s, government agencies started inspections on exports to prevent substandard bicycles from being sent overseas. According to Chu, aside from the control of exported goods, state agencies also started raids aimed to close domestic "underground bicycle factories."[44] It seems likely that this state action arose through a combined concern with maintaining high-quality production and with answering the political lobbying of large manufacturers facing undercutting from such illicit competition.

One of the leading Taiwanese cycling industry companies was Giant, an important part of the favorable public image of Taiwan as a technology-intensive economy.[45] Between its founding in 1972 and 1986, Giant was an OEM for the North American Schwinn, one of the first high-profile cycling industry companies to develop a global supply chain. When the contract with Schwinn was broken in 1986, Giant started to market bicycles under its own brand. At the same time, the company became engaged in a research project that aimed to develop and commercialize CFRP bicycle-frame technology. This project was partially sponsored by the Taiwanese government under its program for the Materials Research Laboratory, a subsidiary of the Industrial Technology Research Institute (ITRI), a government research and development institute established in 1973. ITRI's aim was to develop technologies that would help Taiwanese companies manufacture and export new high-tech products. The Carbon–Fiber Bicycle Development Project was started in 1984, and was jointly funded by the government and several bicycle industry companies.[46] However, other companies withdrew their financial support for the project, and only Giant sponsored it through to success, thus receiving the benefits of sustained investment. In 1987, Giant was able to market the final product of this project. The bicycle—named Giant Cadex 980 C—was designed similarly to the KG-86; the frameset was built with the CFRP tubing jointed with aluminum lugs. But from the start, Giant, contrary to European companies, aimed to manufacture this bike on a larger scale, not as a hand-built product for professionals. As Giant's website claims, this was the "world's first affordable carbon-fiber road bike for the masses."[47] Subsequently, Giant started to release further CFRP framesets under the company's top-end TCR series bicycles. To provide further innovations, the company hired Mike Burrows, a renowned British bicycle engineer who designed several innovative bicycles, such as the Lotus Type 108, a monocoque CFRP bicycle successfully used during the track-cycling competition of the 1992 Barcelona Olympics. Today, Giant is the largest cycling manufacturer in the world and considered one of the few world leaders in technological innovativeness as well as in organizing vertically integrated production systems. Bikeradar, one of the most popular websites dedicated to cycling, regularly publishes "Inside Factory" reportages and, in 2014, published a lengthy piece about Giant's factory, emphasizing the company's prowess in manufacturing of CFRP and organizing highly complex and effective systems for the management of frame assembly.[48]

In the 1990s, several non–Taiwanese bicycle manufacturers started using CFRP in the framesets of top-end racing bicycles. Aside from the introduction of the new material design, the 1990s and 2000s saw another significant change in the industry. To cut costs, European and US companies moved production to Taiwan, where both a cheap labor force and local companies with experience in bicycle manufacturing were available. The necessity of cuttting costs perceived by bicycle manufacturers in the 1990s and 2000s, along with the growing competition and the introduction of new materials, were likely experienced by manufacturers as two different priorities. However, at that time, outsourcing manufacturing to Taiwan was supported by both aims. Companies who decided to move their production to Taiwan were able to cut costs due to cheaper labor forces and, with the know-how of local facilities, also introduce into their operations technical innovation for both producing CFRP and building bicycle frames.

Despite a steady supply of material and know-how on the manufacturing sporting goods with that material, securing Taiwanese subcontractors for CFRP components had a significant marketing disadvantage. Consumer goods with a sticker saying "Made in Taiwan," especially in high-tech sectors, were still perceived by consumers as poorly manufactured products, no matter under which brand those goods might be sold.[49]

In a book on the growing Taiwanese high-technology sector, John A. Mathews and Dung-Song Cho point out frequent accusations of Taiwanese high-tech industries ranging from "piracy of intellectual property, to savage labor practices, to government handouts, to endless imitation of others."⁵⁰ As they explain, most of those accusations, except for those of worker exploitation, came from business competitors rather than from potential customers of Taiwanese products. Further, I will show how consumers express their opinions, including those on the made-in-Taiwan stereotype. Such opinions focus on the quality of manufacturing as well as the business practices of manufacturers who hide the fact that the product was manufactured by subcontractors in Taiwan rather than Italy or France.

Once again referring to Appadurai, we may call such stereotyping one of "peculiarities of knowledge that accompany relatively complex, long distance, intercultural flows of commodities."⁵¹ It is by no means automatic that any particular information about a material or product will become important to those who may purchase or use those commodities, and the focus by Western audiences on the site of manufacture—as remote from and in significant ways unlike their own locale—bears close analysis.

The notion of "intercultural flow" is relevant here, since Western consumers consider products made in East Asia as commodities made by an other and distant culture. From consumers' point of view, the country of origin from which a product derives—in this case, Taiwan—is an important factor in estimating the quality and value of that product. In 1994, Durairaj Maheswaran conducted a research survey on the link between the country of origin of computer components and the perception of quality by eventual customers. Generally, respondents perceived products made in East Asia as less desirable than those manufactured in Europe or the United States.⁵² Maheswaran also quoted free thoughts provided by respondents when asked about their opinion on specific countries. Thoughts such as "German products have great engineering" and "Thailand workers do not have good technical skills" clearly show how COO stereotypes contributed to notions of nations' technical ingenuity, or lack thereof.⁵³ Spokesperson for Giant Jeffrey Sheu, in an interview, claimed that perceptions of the low quality of made-in-Taiwan products constituted one of the most significant constraints to entering global markets. "Converting the stereotype of MIT [made-in-Taiwan] products became our first challenge," said Sheu. "Our response was to specialize in high-end bicycles, a strategy to overturn all previous assumptions about MIT products."[54]

Crucial to our understanding of this choice of strategy was that CFRP is primarily identified as a Taiwanese product; a Taiwan-made bicycle using this material and accepted by cyclists would be, advocates hoped, indisputably a credit to Taiwanese heritage. But cyclists' COO expectations were not so easily trumped, as shown in a highly controversial statement by one of the most prominent figures in professional cycling. In 2009, Pat McQuaid, the president of the Union Cycliste Internationale–International Cycling Union (UCI), the ruling body of professional cycling, delivered an unofficial speech in which he accused Asian manufacturers of the overpricing of poorly manufactured frames and criticized the whole innovation trend. This statement was reported by a trade journal: "Quaid had said the bike industry was 'turning out thousands of carbon fiber frames, at a cost of maybe $30 or $40 apiece, and that same bike is ultimately sold as a bike for five or six thousand Euros.'" The journal added, "McQuaid also claimed Asian-made frames, unlike steel frames of old, are unsafe and cause crashes because they are light, guilty of 'hopping all over the place.'"[55] McQuaid's claim, which was protested by bicycle manufacturers, shows how one of the most important figures in professional cycling formulated concern over the possible lack of safety as a side effect of the introduction of new material. In this opinion, the lightness of a bicycle made with CFRP is viewed not as a factor that significantly improves an athlete's performance, but as one that potentially causes an unstable bicycle in contrast to older and safer metal-alloy frames. At the same time, this innovative industry is still burdened with the made-in-Taiwan stereotypes. The Taiwanese bike makers' dismay over McQuaid's claim was worsened because it was not an isolated opinion. Rather, he shared a belief in an evaluative and subjective COO stereotype highly critical of Asian producers that had become popular on cycling internet forums. McQuaid's influential elite voice and more grassroots complaints were mutually supporting.

WHO MADE YOUR BIKE? EUROPEAN MANUFACTURERS AND THE "MADE IN TAIWAN" LABEL

In the 1990s, along with the growing popularity of healthy lifestyles and fitness, cycling evolved from a niche discipline into a lifestyle sport. Subsequently, the racing bicycle evolved from merely a piece of sports equipment into a more "lifestyle object." This evolution was fueled by the bicycle companies, which included several new marketing strategies for wider audiences and not only for cycling aficionados.[56] CFRP framesets that enabled multiple new designs provided manufacturers with the possibility of using a strategy to differentiate product lines and audiences. Some contemporary, more flamboyant CFRP component designs are apparently made to be eye-catching rather than made to enhance any performance benefits. In such a context, as contemporary marketing practices in the industry show, the often-stigmatizing sticker "Made in Taiwan" does not fit well on a bicycle with a price tag of $6,000 to $10,000 offered by a high-profile Italian or French company. In this section, I show how European cycling companies are marketing CFRP bicycles made by Taiwanese subcontractors while trying to keep the traditional brand values. Such practices have been widely discussed in cycling industry magazines and by high-profile business outlets such as CNBC.[57] First, I discuss how bicycle manufacturers attempted to merge traditional artisanal values with material innovativeness, and second, I discuss how bicycle users in turn challenged manufacturers' claims.

High-profile companies such as the aforementioned Colango, Pinarello (started in 1952), and Bianchi (dating all the way back to 1885) are manufacturers currently identified as producers of several successful high-end "Made in Italy" bikes. Colnago is referred to as the "most famous global icon of Italy's premium bike-making heritage."[58] Colnago built its marketing strategy on three focal points: artisanal production under the direct supervision of "master frame builder" Ernesto Colnago, the use of a premium material, and premium Italian design. Official statements claim that the company is innovative by being a first-mover in the CFRP frames: "Colnago was the first frame builder to see the potential of carbon fibre in the evolution of the cycling." And yet, much is also made of the ways in which the firm still respects traditional values: "Colnago takes age-old craftsmanship into the future."[59] In the 1990s, CFRP frames built by Colnago were made in Italy with material provided by Italian supplier ATR, a company that also supplies European aerospace and super-car industries.[60] In the 2000s, however, Colnago joined A-Team, a Taiwan-based sourcing and product-promotion group. A-Team includes members Giant, Merida, and other Taiwan-based component makers. A US industry journal article of 2005 claims that Colnago, at that time, sold twenty-three hundred bikes in the US per year, and that seven hundred of these bikes would come from Taiwan.[61] In an official statement from Ernesto Colnago, we can see how the company emphasizes that high-end CFRP bikes are still made in Italy and with Italian material: "For the 2006 model year, Colnago will be sourcing two entry-level aluminum road bike models from Giant, made to Colnago's spec and frame geometry and for sale in Europe and Asia only. All other Colnago bicycles are made in Italy.... No Colnago carbon fiber frames are made at Giant and none will be, as Colnago has a long-term sourcing agreement in place with ATR for carbon fiber bicycle frames."[62] Colnago goes on to reinforce his listeners' comprehension regarding the company's understanding of a stratified world of bike production and consumption: "Let me be completely clear; all Colnago bicycles will be designed and engineered in our Cambiago, Italy headquarters as always. But starting later in 2005, we will have two different production sources for Colnago bicycles. ALL of our high-end bicycles will be made in Italy as they have been since 1954. All of our mid-range bicycles will be designed in Italy and produced in Taiwan."[63] It is evidently of importance that no one mistake Colnago's intentions, or suspect the company of hiding what are seen by speaker and audience alike as real differentials in the value of goods made in different places.

Not surprisingly, this is also a topic widely discussed among the internet cycling communities, thus showing the consumer reception of companies' strategies and the perception of artifacts with "Made in Italy" and "Made in Taiwan" labels.[64] Currently, not only cycling magazines but also several blogs dedicated to cycling and bicycle design play a role as cultural intermediaries in this sporting-goods market. Quotations used here came from high-profile blogs on which readers extensively commented, and links to those blog entries were brought into several forum discussions. Quotations from forum discussions came from the widely read Road Bike Review website. The cycling bloggers and forum members frequently discuss such manufacturers' practices, with the aim of verifying the country of origin of a bike that they own or intend to purchase. There is a whole genre of forum topics and blog entries focused on "who actually made your bike," where information of varied reliability and pieces of gossip are shared.[65] For instance, one popular blog lists Italian, French, and US OEM Taiwanese companies, apparently to alert others to what we might call the "mixed heritage" of the machines reaching Western markets.[66]

This is not a universe of judgments easily labeled as nationalist or racist, however. The hobbyists' circulation of facts and opinions regarding bicycle or parts production reflects complex concerns about relationships between manufacturers and consumers, on top of which notions of nationality are layered. When I started my research, I expected to find several entries with claims regarding the poor quality of bicycles made in Taiwan. Actually, I have instead found entries that criticize manufacturers' practices of withholding information on the actual country of origin, but not the quality of products manufactured in Taiwan. Here is a typical complaint from the Road Bike Review forum discussion: "There's one gossip I heard that '08 Bianchi frame are made in Taiwan, even RC [Reparto Corse, the company's flagship model] one. None of hand-made craft work on the legendary production anymore. I don't mean Taiwan's manufacturing is poor, but just want to make sure and curious the reason."[67] A more elaborate entry on the Lovely Bicycle! blog provides a good explanation of the fascination with Italian or French bicycles, rather than Taiwanese: "Personally, I prefer it when a bicycle is made as part of a small production run, by hand, and within a culture that I have some personal connection to. It's just more interesting to me that way. But I have nothing against Taiwan or China per se, as long as the specific factory provides good working conditions, employs environmentally safe practices and uses high quality methods of production."[68] This entry shows that a racing bicycle could be perceived by users as a product of a specific culture rather than the result of a manufacturing process only. Buying bikes made in countries with more than a hundred-year-old tradition of cycling and bicycle manufacturing may provide users with cultural references that could be used in shaping social identity as someone who has some established links to this culture, such as, for instance, a connoisseur of the "Italian style." But crucially, that entry also suggests that multiple values, including labor and environmental values of a country of origin, are at least worthy of invocation, if not actually driving consumption choices. It is important to note that such values are not neutral, since even evaluations of work conditions and attention to quality can be deeply interwoven with some nationalistic or racist stereotypes, as illustrated above with Maheswaran's research.

I have found several entries that show another notion of a contemporary bicycle as an object primarily of scientific and engineering practices, not an object with some explicit cultural values, as aforementioned statements include. On the one hand, on popular internet forums, one can find regular claims that intended buyers are concerned with the fact that their bicycle frame will be physically manufactured in a different culture and only later transferred into their own culture by providing it with a legitimate logo and other markings of an Italian brand. We have to remember, however, that such concerns should be considered not solely identity-neutral consumers' concerns about the possible dishonesty of manufacturers. They, rather, should be interpreted within the politics of national and racial identity brought to light, above.

On the other hand, on same forums, one can find claims that when it comes to products made with CFRP, the made-in-Taiwan stigma is fading away, since those are high-tech products that should be evaluated through information about the science that goes into production rather than any artisanal skills that might be involved. In this discourse, a manufacturer's reasearch and development budget is more relevant to estimating a bicycle's value than is the manufacturer's national culture in any simple sense: "I would rather leave carbon composite engineering to the big boys with R&D budgets and steel, aluminum and bamboo to the artisans. The pace is such that it's beyond their reach now."[69] This stress on practicality and resources, however, still associates different kinds of manufacturing settings with particular priorities. Different locales can be reliably assumed, this entry suggests, to bring different approaches to bicycle design and manufacture.

Their nuances notwithstanding, the quotations above together show a shift in the perception of bicycles from an object made by an artisan into an object designed within the framework of technoscience and manufactured in a plant located in a technoscientific nexus. I argue that the introduction of CFRP is crucial in the perception of the legitimate, high-end contemporary racing bicycle. The bicycle made with this "space age material" and designed with CAD software, taking into account its aerodynamic profile, is not a product of artisanal production anymore. To use Bijker's term, it is a product of a different "technological frame" based on industrial-scale composite manufacturing rather than artisanal metal-alloy manufacturing.[70] Currently, the desirable bicycle for aficionados and accomplished athletes is more and more often perceived as a product of a global high-tech industry; the country of origin is not of primary relevance if a manufacturer is seen to have secured a high research and development budget and a reliable manufacturing process. Cycling journalist Carlton Reid, in his reportage on bicycle industries in Asia, shows how the change of the "technological frame" impacted the perception of the manufacturing process environment: "Purists argue that high-end bicycles are not supposed to be made in China, they are meant to be hand-crafted by chain-smoking artisans, welding gossamer-thin steel tubes in dank workshops, all for love not money. But carbon's not like that. Composite bike frames are made in sterile, well-lit labs."[71] Reid referred to China since several companies are currently moving manufacturing from Taiwan to mainland China, where a much cheaper labor force is available. However, this juxtaposition of chain-smoking "master frame builders," as they were referred to during the twentieth century, and the engineers and factory employees toiling in clean twenty-first century labs, is also relevant to Taiwan. During the last three decades, Taiwanese companies firmly secured their position as manufacturers of electronic components where both extended research-and-development resources and a high level of precision and cleanliness are required.

CONCLUSION

In this chapter, I have shown how the introduction of a composite "wonder material" in the cycling industry had an impact on cultural values of this industry based on conceptions of artisanal production and metallurgy. First, I discussed how, during the course of the twentieth century, high-end racing bicycles built by European manufacturers were marketed with claims on such bases as the established handicraft in metal-alloy manufacturing and the technological ingenuity of Italy and France. Further, I showed how the technology of manufacturing bicycles with CFRP has been developed in Taiwan, and currently most high-end bicycle components are manufactured there. European companies moved their manufacturing to Taiwan, as well, while attempting to market those products with stickers indicating that they were "Made in Italy" and "Made in France," thus avoiding the "Made in Taiwan" label that still risks, apparently, invoking stereotypes concerning the lack of technological ingenuity and reliability in Asia.

On the one hand, in the world of bicycle production and consumption, CFPR was viewed widely from its inception not only as material that brings several apparent benefits to athletes' performance, but also as a product of contemporary science. On the other hand, as quoted discussions show, the introduction of this material demonstrates that technical knowledge is perceived as not universal; rather, such knowledge (and its applications) are embedded with country-of-origin stereotypes. In the twenty-first century, pragmatic potentates in the cycling industry moved their production to Taiwan, while at the same time attempting to keep their core brand values. They have aimed to show that a bicycle made with CFPR is still a product of European technological ingenuity.

It is striking to see how two different "technological frames" somehow coexist when it comes to the evaluation of high-end racing bicycle frames. Consumers have expressed their concern regarding in which country, or rather, in which culture, this particular product is made. But at the same time, we can find on both internet forums and in media discourse on cycling statements expressing that science prevails over heritage: If this frame is made with the same wonder material that was used to build a space shuttle or the B-2 Spirit bomber, one should put aside geographic origin to consider instead whether or not a manufacturer has had enough of a knowledge base and enough financial resources to carry out proper research and development of this product. We can perhaps imagine multiple logics that lead the contemporary bicycle owner to make a purchase: Shall I believe that my bicycle, an object in which most components are made with CFRP, is worth, say, $8,000, as priced, because it is lovingly made or because scientists helped make it? Could both arguments support the purchase, or do the artisan and scientist remain mutually exclusive figures of influence in contemporary industrial cultures? This case study shows how many factors come into play during human engagements with a new material when manufacturers release consumer products made with wonder-inducing material—one known to have been used recently to build, for example, technical marvels in the aerospace industries. In this chapter, I have outlined several largely undisputed advantages that CFRP has over metal alloy as a material for racing-bicycle frames and other components. However, this is definitely not a case (if such a case can even exist) where a new material simply replaces an old one solely because of its physical properties. This case, rather, shows how an introduction of material innovation could interact with the reproduction of national identities, and with—relatedly—ideas about what exactly might constitute a national technological achievement.

ACKNOWLEDGEMENTS

I would like to express my gratitude to Amy Slaton and two anonymous reviewers, who provided me with a lot of insightful comments and suggestions that significantly helped to improve this essay.

NOTES

1. For an outline of CFRP properties, see Khalid Lafti and Maurice A. Wright, "Carbon Fibers," in *Handbook of Composites*, 2nd ed., ed. S. T. Peters (London: Chapman & Hall, 1998), 169–201.
2. Mimi Sheller, *Aluminum Dreams. The Making of Light Modernity* (Cambridge, MA, and London: MIT Press); Quentin R. Skrabec, *Aluminum in America: A History* (Jefferson: McFarland and Company, 2017).

3. General advantages of the CFRP option over metal-alloy use in racing-bicycle construction has been discussed in Michael Kaiser and Norbert Himmel, "Carbon Fiber Reinforced Plastics—Trendsetting Material for High Performance Racing Bike Chassis," in *The Engineering of Sport 6*, vol 3: *Developments for Innovation*, eds. Eckehard Moritz and Steve Haake (New York: Springer, 2006), 123–28. For a general discussion on the bicycle as a human-powered technology, see David Gordon Wilson, *Bicycling Science* (Cambridge, MA, and London: MIT Press, 2004).
4. Aleksandar J. Subic and Steve J. Haake, eds., *The Engineering of Sport: Research, Development and, Innovation* (Oxford: Wiley-Blackwell, 2000), 3–4.
5. This case refers to a broader phenomenon of the perception of technological progress as a product of particular cultures. For instance, Stanford L. Moskowitz, while discussing innovations in material technologies, refers to most innovations as particularly American or European achievements. Sanford L. Moskowitz, *The Advanced Materials Revolution: Technology and Economic Growth in the Age of Globalization* (Hoboken, NJ: Wiley, 2009).
6. The role of the so-named Industrial Revolution in nation-building processes in Europe is discussed in Lars Magnusson, *Nation, State and the Industrial Revolution: The Visible Hand* (London: Taylor & Francis, 2009).
7. Benedict R. Anderson, *Imagined Communities: Reflections on the Origin and Spread of Nationalism* (London: Verso, [1983] 1991).
8. Thomas Zeller, *Driving Germany: The Landscape of the German Autobahn, 1930–1970* (Oxford and New York: Berghahn Books, 2010).
9. Tiago Saraiva, *Fascist Pigs: Technoscientific Organisms and the History of Fascism* (Cambridge, MA, and London: MIT Press, 2016).
10. For an overview of research trends on COO stereotypes, see Nicolas Papadopoulos and Louise A. Heslop, eds., *Product-Country Images: Impact and Role in International Marketing* (Binghamton, NY: International Business Press, 1993); and Warren J. Bilkey and Erik Nes, "Country-of-Origin Effects on Product Evaluations," *Journal of International Business Studies* (Spring/Summer 1982): 89–99; Michael Chattalas, Thomas Kramer, and Hirokazu Takada, "The Impact of National Stereotypes on the Country of Origin Effect: A Conceptual Framework," *International Marketing Review* 25, no. 1 (2008): 54–74.
11. Arjun Appadurai, "Introduction: Commodities and the Politics of Value," in *The Social Life of Things: Commodities in Cultural Perspective*, ed. Arjun Appadurai (Cambridge: Cambridge University Press, 1986), 3–63 and 41.
12. Chattalas, in his doctoral dissertation, provides an extensive review of a large number of such studies: Michael J. Chattalas, "The Effects of National

Stereotypes on Country of Origin-Based Product Evaluations" (PhD diss., City University of New York, 2005).

13. David V. Herlihy, *Bicycle: The History* (New Haven and London: Yale University Press, 2004), 376–402.
14. Winfried Wolf insightfully grasps the shift from railways to cars as symbols of technological prowess: Winfried Wolf, *Car Mania: A Critical History of Transport* (London and Chicago: Pluto Press, 1996), 67–90.
15. David Wilson provides an in-depth analysis of how speed is a simultaneous result of the efficiency of a cyclist and bicycle design: David Gordon Wilson, *Bicycling Science* (Cambridge, MA, and London: MIT Press, 2004).
16. Hugh Dauncey and Geoff Hare, eds., *The Tour de France 1903–2003: A Century of Sporting Structures, Meanings and Values* (London and Portland, OR: Frank Cass, 2005), 3–4.
17. Roland Barthes, "The Tour de France as Epic," *The Eiffel Tower and Other Mythologies* (Los Angeles: University of California Press, 1997). Originally published in Roland Barthes, *Mythologies* (Paris: Seuil, 1957).
18. For the history of the Giro d'Italia, see Bill McGann and Carol McGann, *The Story of the Giro D'Italia*, vol. 1: *A Year-by-Year History of the Tour of Italy* (Ozarks: McGann Publishing, 2011); John Foot, *Pedalare, Pedalare: A History of Italian Cycling* (London: Bloomsbury, 2011).
19. Wiebe E. Bijker, *Of Bicycles, Bakelites, and Bulbs: Toward a Theory of Sociotechnical Change* (Cambridge, MA, and London: MIT Press, 1995), 19–100. Cf. O. A. van Nierop, A. C. M. Blankendaal, and C. J. Overbeeke, "The Evolution of the Bicycle: A Dynamic Systems Approach," *Journal of Design History* 10, no. 3 (1997): 253–67.
20. Appadurai, "Introduction," 41.
21. Bijker, *Of Bicycles, Bakelites and Bulbs*, 96.
22. Wilson, *Bicyling Science*, 32.
23. Bruce D. Epperson, *Peddling Bicycles to America: The Rise of an Industry* (Jefferson and London: McFarland, 2010), 85.
24. Epperson, *Peddling Bicycles*, 110.
25. Velo Base, "Home" (website), accessed February 10, 2015, http://velobase.com/default.aspx.
26. William Fotheringham, *Cyclopedia: It's All About the Bike* (Chicago, IL: Chicago Review Press, 2011), Kindle edition.
27. For a history of innovation in bicycle derailleur systems, see Frank J. Berto, *The Dancing Chain: History and Development of the Derailleur Bicycle*, 4th ed. (San Francisco, CA: Van Der Plas Publications, 2012).

28. For instance, the Italian Ways website, which promotes Italian national identity, includes an elaborate biography of Ugo de Rosa, the founder of the de Rosa company and one of the most notable frame builders, next to stories of the origins of pizza and Renaissance sculptors; Ron Miriello, "Ugo de Rosa, master frame builder," Italian Ways, December 15, 2013, http://www.italianways.com/ugo-de-rosa-master-frame-builder/.
29. James B. Twitchell, *Living It Up: Our Love Affair with Luxury* (New York: Columbia University Press, 2002), 126.
30. Mark Tungate offers a historical discussion on the origins of several luxury commodities: Mark Tungate, *Luxury World: The Past, Present and Future of Luxury Brands* (London and Philadelphia, PA: Kogan Page, 2009).
31. For a brief history of Italian industrial design, see Grace Lees-Maffei and Kjetil Fallan, eds., *Made in Italy: Rethinking a Century of Italian Design* (London and New York: Bloomsbury, 2014).
32. Paul Rosen, "The Social Construction of Mountain Bikes: Technology and Postmodernity in the Cycle Industry," *Social Studies of Science* 23, no. 3 (1993): 479–513.
33. Sheller, *Aluminum Dreams*, 85–113.
34. Dave Atkinson, "125 years of Mavic: A Ride through Cycling History," Road.cc, May 14, 2014, http://road.cc/content/feature/118858-125-years-mavic-ride-through-cycling-history.
35. Eric Schatzberg offers an interesting discussion on the role of politics in the application of aluminum in aircraft design: Eric Schatzberg, "Ideology and Technical Choice: The Decline of the Wooden Airplane in the United States, 1920–1945," *Technology and Culture* 35, no. 1 (1994): 34–69.
36. D. M. Peters, "The New Steel: Carbon-Fibre-Reinforced Plastics: The Materials That May Revolutionize Aircraft Design," *FLIGHT International*, October 24, 1968, 669–70; Marcus Langley, "Carbon Fibres—The First Five Years," *FLIGHT International*, September 9, 1971, 406–8.
37. E. F. Lindsley, "Composites—How They'll Make Cars Lighter, Stronger," *Popular Science*, April 1978, 89.
38. Elizabeth Shove, Matthew Watson, Martin Hand, and Jack Ingram, "The Materials of Material Culture: Plastic," *Design of Everyday Life* (Oxford and New York: BERG, 2007), 93–116. See, also, Jeffrey L. Meikle, *American Plastic: A Cultural History* (New Brunswick, NJ: Rutgers University Press, 1995).
39. Carlos Tsai, "Carbon Fibre Valley in Taiwan: Global Financial Tsunami has Lead Taiwan's CFRP Industry upward," Sports, accessed April 15, 2015, http://www.sports.org.tw/e/report/2010yearbook_subject/01carbon.html.

40. For a general discussion on the CFRP in sporting goods, see Subic and Haake, *The Engineering of Sport*.
41. Look, "Home" (website), accessed April 15, 2015, http://www.lookcycle.com/en/us/look-cycle/histoire.html.
42. Shelley Rigger, *Why Taiwan Matters: Small Island, Global Powerhouse* (Lanham, MD: Rowman & Littlefield, 2011), 76, Mobi edition.
43. For the discussion on the role of the plastics industry in Taiwanese economy, see Tony Yu Fu-Lai, *Entrepreneurship and Taiwan's Economic Dynamics* (Heidelberg and New York: Springer, 2012); Alice H. Amsden and Wan-wen Chu, *Beyond Late Development: Taiwan's Upgrading Policies* (Cambridge, MA, and London: MIT Press, 2003).
44. Wan-wen Chu, "Causes of Growth: A Study of Taiwan's Bicycle Industry," *Cambridge Journal of Economics* 21, no. 1 (1997): 57. Simon Partner in his study of the Japanese consumer-electronic industry discusses the poor reception of Japanese electronics in the US market in the 1950s, and shows how Japanese manufacturers and the Ministry of International Trade and Industry (MITI) actively challenged negative "Made in Japan" stereotypes in the 1960s; Simon Partner, *Assembled in Japan: Electrical Goods and the Making of the Japanese Consumer* (Berkeley: University of California Press, 1999).
45. For an overview of Giant as a Taiwanese high-technology success story, see Weld Royal, "Made in Taiwan," *Industry Week,* February 15, 1999, 58, 62.
46. Yu-Shan Chen et al., "Technological Innovations and Industry Clustering in the Bicycle Industry in Taiwan," *Technology in Society* 31, no. 3 (Aug. 2009): 207–17; Amsden and Chu, *Beyond Late Development*, 86.
47. Giant, "1987: Carbon Fibre for the Masses," accessed April 15, 2015, http://www.giant-bicycles.com/en-au/aboutgiant/ourhistory/item/carbon.fiber.for.the.masses/3/.
48. James Huang, "Inside Giant's Taiwan Frame Factory," Bike Radar, May 2, 2014, http://www.bikeradar.com/road/gear/article/inside-giants-taiwan-frame-factory-part-one-39835/.
49. For a discussion of attempts at challenging negative connotations of "Made in Taiwan" COO stereotyping through advertising, see Beth Snyder Bulik, "Made in Taiwan," *Advertising Age*, June 20, 2005, 1, 37.
50. John A. Mathews and Dong-Sung Cho, *Tiger Technology: The Creation of a Semiconductor Industry in East Asia* (Cambridge: Cambridge University Press, 2000), xiv.
51. Appadurai, "Introduction," 41.
52. Durairaj Maheswaran, "Country of Origin as a Stereotype: Effects of

Consumer Expertise and Attribute Strength on Product Evaluations," *Journal of Consumer Research* 21, no. 2 (1994): 354–65.
53. Maheswaran, "Country of Origin," 358.
54. Elaine Hou, "Giant More Than a Bicycle Brand," *Taiwan Today*, October 22, 2010, http://taiwantoday.tw/ct.asp?xItem=124520&CtNode=1743.
55. Carlton Reid, "Pat McQuaid Slammed by Industry Execs over 'Not True' Carbon Comments," Bike Biz, September 2, 2011, http://www.bikebiz.com/news/read/pat-mcquaid-slammed-by-industry-execs-over-carbon-critique/011819.
56. Paul Rosen discusses the development of a mountain bike as lifestyle object: Rosen, "The Social Construction of Mountain Bikes."
57. Carlton Reid, "A Bicycle Made by Two," CNBC Business, September 2008, http://www.cnbcmagazine.com/story/a-bicycle-made-by-two/503/1/.
58. Doug McClellan, "Colnago Resurgent in U.S.," Bicycle Retailer & Industry News, February 9, 2012, http://www.bicycleretailer.com/north-america/2012/02/09/colnago-resurgent-us.
59. Colnago, "Materials" accessed April 15, 2015, http://colnago.com/materiali-2/.
60. ATR Group, "Home," accessed October 10, 2012, http://www.atrgroup.it/.
61. John Crenshaw, "Colnago Rebutts Sourcing Reports," Bicycle Retailer and Industry News, August 15, 2005, http://business.highbeam.com/103/article-1G1-135579315/colnago-rebutts-sourcing-reports.
62. Crenshaw, "Colnago Rebutts."
63. "Colnago's Entry Level Bikes to Be Produced in Taiwan," Bike Biz, March 1, 2005, http://www.bikebiz.com/news/read/colnago-s-entry-level-bikes-to-be-produced-in-taiwan/02071.
64. The role of consumers in the social construction of technology is discussed in Nelly Oudshoorn and Trevor Pinch, eds., *How Users Matter: The Co-Construction of Users and Technology* (Cambridge, MA, and London: MIT Press, 2003).
65. For instance, see Lovely Bicycle, "Where Was Your Bicycle Made? . . . and Does It Matter?" (blog), November 14, 2011, http://lovelybike.blogspot.com/2011/11/where-was-your-bicycle-made-and-does-it.html; The Inner Ring, "Who Made Your Bike," February 19, 2012, http://inrng.com/2012/02/who-made-your-bike/.
66. Memoirs on a Rainy Day, "OEM Taiwanese Carbon Fiber Bike Companies" (blog), September 3, 2009, http://range.wordpress.com/2009/09/03/oem-taiwanese-carbon-fiber-bike-companies/.
67. Road Bike Review, "Is It True '08 Bianchi Frames Are Made in Taiwan??" (online forum), last modified August 1, 2008, http://forums.roadbikereview.com/bianchi/true-08-bianchi-frames-made-taiwan-135765.html.

68. Lovely Bicycle, "Where Was Your Bicycle Made."
69. The Inner Ring, blog comment by Rider Council, February 19, 2012, http://inrng.com/2012/02/who-made-your-bike/.
70. Bijker, *Of Bicycles, Bakelites, and Bulbs*, 122–27.
71. Carlton Reid, "A Bicycle Made by Two."

CHAPTER EIGHT

GRENFELL CLOTH

Rafico Ruiz

> *A medium of communication has an important influence on the dissemination of knowledge over space and over time and it becomes necessary to study its characteristics in order to appraise its influence in its cultural setting.*
> —Harold Innis, "The Bias of Communication"

I've only *touched* Grenfell Cloth once (see figure 1). It was at the provincial archives of Newfoundland and Labrador, and the array of colored cloth samples came attached to a small piece of cardboard. The cloth itself is both rough and silky, and is hard to the touch. It feels waterproof, made to last against the wind. In many ways, it is a cloth of the elements, with wind, water, cold, and heat all shaping its ultimate lightness of weight, durability, and sense of rugged materiality. It was created to embody a particular relationship between humans and their environments that could account for the latter's extremes.

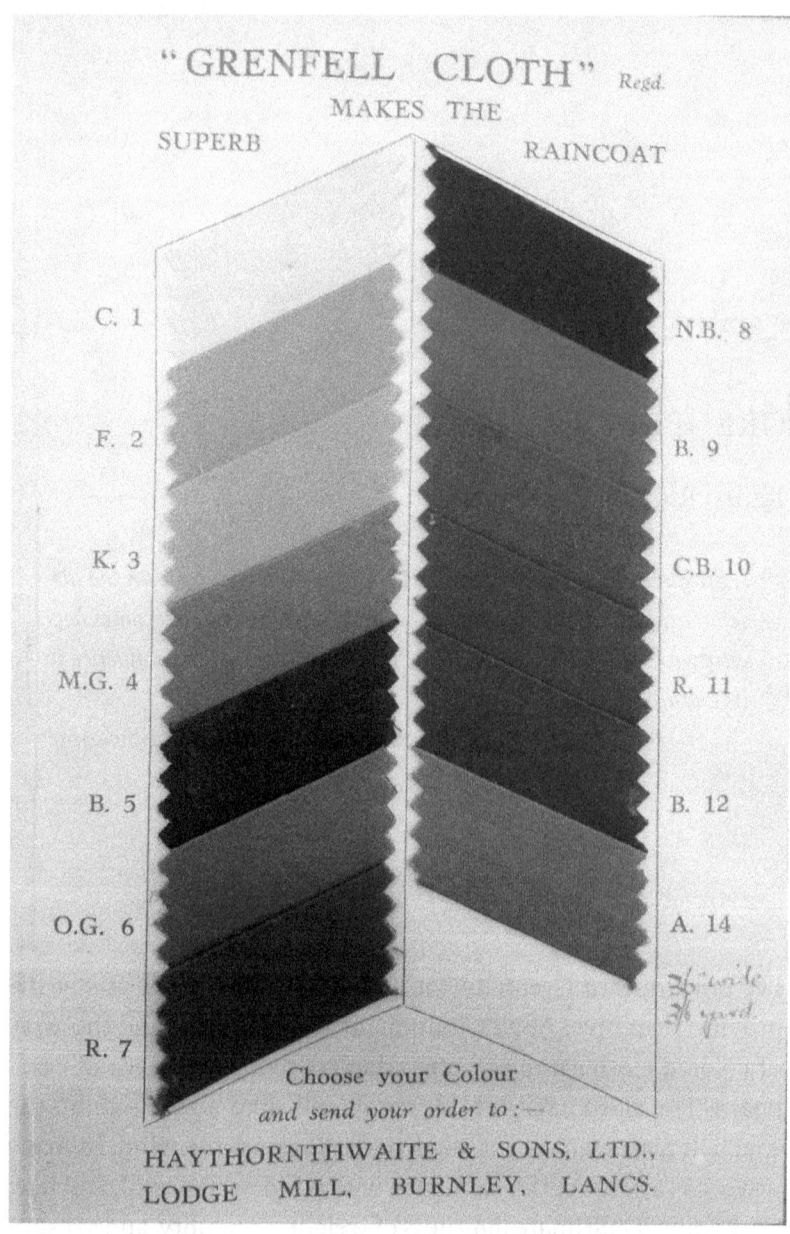

Figure 1. Grenfell Cloth samples, ca. 1930. Provincial Archives of Newfoundland and Labrador. *Source*: photo courtesy of the author.

Figure 2. Grenfell Cloth garment, ca. 1930, Grenfell Interpretation Center, St. Anthony, Newfoundland and Labrador. *Source*: photo courtesy of the author.

I *saw* a garment made of Grenfell Cloth only once as well (see figure 2). It was behind glass at the Grenfell Interpretation Centre in St. Anthony. It had a strange durability to it, a quality of aged cotton, a natural fiber, that seemed to give it an archeological tone.

In historiographical terms, I came to know of Grenfell Cloth through its archival recasting in the holdings of the International Grenfell Association. There "it" was, in a series of manila folders that could unfold its documentary and artefactual story (see figures 3 and 4). In pamphlets, advertising sheets, samples of the cloth itself, and correspondence, Grenfell cloth's artefactual work was marked. Its genesis story is one of multiple agencies and mediating processes, many that are primarily discursive, others political-economic, and yet others that are embodied—with all of these being historic. So, the question presents itself: How does one get a hand on Grenfell Cloth?

*

On December 9, 1936, at the Dorchester Hotel in London, the Duke and Duchess of York hosted The Labrador Ball. A fundraiser in aid of the Grenfell Association of Great Britain and Ireland, the ball also featured a Miniature Labrador Exhibition arranged by the Hudson Bay Company (HBC). The exhibition made up a representative collection of furs from Labrador, salmon caught off the coast of Newfoundland and transported to London via the HBC's then-novel method of quick freezing, as well as an illuminated display of Grenfell Cloth and paintings of Labrador by Rhoda Dawson. Kicking off at 10 p.m., with a welcome by the Countess of Shrewsbury, the charitable evening would continue on with supper, a cabaret performance, and dancing. Tucked away at the back of the ball's program is a full-page ad with a somewhat odd photograph and caption: "The highest front door on earth" (see figure 6). The ad's copy reads:

Figure 3. Promotional material for Grenfell Cloth, ca. 1930. Provincial Archives of Newfoundland and Labrador. *Source*: photo courtesy of the author.

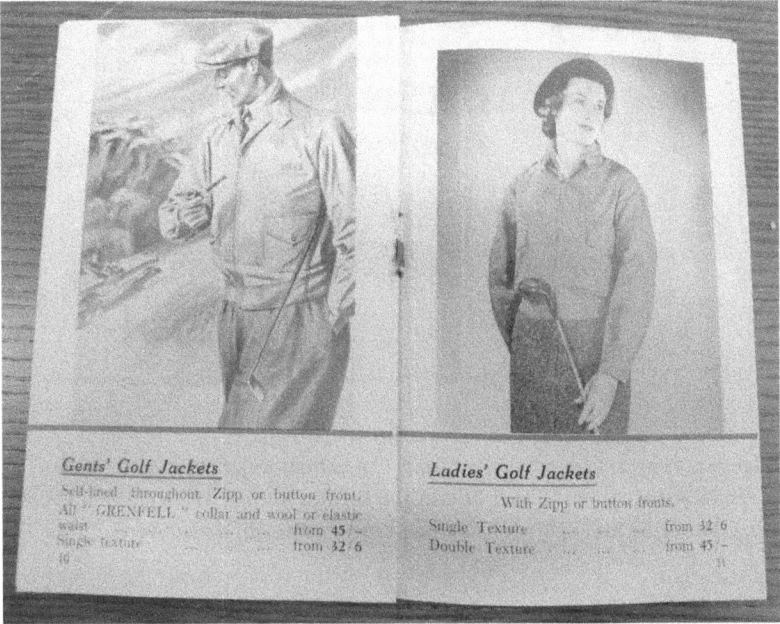

Figure 4. Promotional material for Grenfell Cloth, ca. 1930. Provincial Archives of Newfoundland and Labrador. *Source*: photo courtesy of the author.

Figure 5. Program advertisement from The Labrador Ball: In Aid of The Grenfell Association of Great Britain and Ireland (1936). Provincial Archives of Newfoundland and Labrador. *Source*: photo courtesy of the author.

The highest habitation that man has ever built on this planet was a small Grenfell Cloth tent pitched on the upper reaches of Mount Everest . . . 27,400 feet above sea-level. Scientists were doubtful that a man could sleep at such an altitude and waken again. F.S. Smythe slept for thirteen hours in this little Grenfell tent whilst a blizzard blew outside.[1]

"Grenfell Cloth," as the textile came to be called by the mid-1930s, was a known commodity to attendees of the ball. Just a few years prior, Dr. Wilfred Grenfell, a Protestant medical missionary ministering to the "fisherfolk" in the colony of Newfoundland, himself had sent the Prince of Wales a sample golf jacket made of the cloth. As the figure in the ad intimates, Grenfell Cloth was meant to bridge the gaps between imaginaries of metropolitan weather and the extreme durability required in harsh and remote climates while also a "known" and reliable commodity in the empire's self-fashioning of commercial production.

The story of the invention, production, circulation, and ultimate redeployment of Grenfell Cloth later in the twentieth century picks up on many common and recurring aspects of the Grenfell Mission story; from the introduction of "new" materialities into particular social sites of natural resource engagement to the associative work that immaterial ideologies such as those surrounding cooperative finance could engender. Grenfell began traveling to the outports along the coasts of northern Newfoundland and Labrador in 1892 aboard the medical ship *Albert* sent by the Royal National Mission to Deep Sea Fishermen (RNMDSF). The mission he worked to establish, culminating in the incorporation of the International Grenfell Association (IGA) in 1914, was an organization that would eventually oversee the construction and functioning of hospitals, nursing stations, schools, orphanages, cooperative stores, and light industries, among other institutional types, becoming a vast northern health network that the IGA ran until, in 1981, it was finally transferred over to provincial control.

What is unique about the cloth's particular agential mediation that I will trace here is that it opens up questions that pertain to both the customary historiographical work tracing its particular process of materialization, and suggests how materialist communicative practices can be read back into and onto configurations of historical artifactuality. Lisa Gitelman claims, in relation to conventional media, that "people learn about the past through them, and using media also involves implicit encounters with the past that produced the representations in question."[2] It could thus follow that recasting the knowability and artifactual legibility of a material such as Grenfell Cloth can go some way toward thinking about the meeting point between new materialist ontologies that are reconceptualizing the limits of agency in relation to material processes, and media history as historiography that would do well to acknowledge and incorporate (rather than elide) the performative work of recasting, whether through archeological or technical historicity, the objects at the center of their inquiry. In this sense, such artifacts, whether remediated through the digital or apprehended in the flesh behind display-case glass, are always already made, as well as being always already-evolving processes of historical mediation and recurrences through their discursive framings and very material renderings.

This is an approach that echoes Tiago Saraiva's examination in this volume of the processual story behind the standardization of citrus in Southern California during the Progressive Era. Citrus varieties, for Saraiva, were the result of modernizing practices, such as cloning, and they marked historical boundaries between social classes. These varieties were mobile manifestations of material developments that bound together and blurred the subjects and objects of agential causation. One such mediating phenomena and practice that the Mission brought into northern Newfoundland and Labrador, and yet also brought out into a circulating imperial environment, was the textile known as Grenfell Cloth.

It was during Grenfell's participation in the First World War that he observed the need for a waterproof uniform for soldiers, and found an origin for such garments in Labrador's fishing industry. The result was Grenfell Cloth, an experimental, cotton-based weatherproof textile made in the Haythornwaite mills of Lancashire. As the print publication ad shows, the textile quickly became an integral technology to the earliest "conquest" expeditions of Mount Everest in the 1930s, yet it was also used in setting land and water-speed records in the same decade, and achieved considerable success in the British outdoor clothing market. Grenfell Cloth has survived across the twentieth century, and is today a relatively enduring heritage brand and commodity in the global textile market.

Here, I would like to place Grenfell Cloth into what Katie King calls a "transmedia story." For King, a transmedia story is a performative media ecology that can create knowledges through the ways in which it is constructed as an object of inquiry.[3] Yet, it is also a generative strategy that recognizes a hermeneutic hand in the configurations of historical artifactuality. In King's estimation, these sorts of strategies can give onto "transdisciplinary knowledges" that work toward posthumanities approaches to agency (now distributed across multiple media platforms, fields of intentionality, and artifactualities) that recognize the limited control writers (and their writing technologies) hold in the making of these agential environments. As such, the goal of this placement is to describe the mediated life of a textile and the ways in which historiographical practices are being shaped by such processes of mediated knowledge production. I want to ask: How can we apprehend Grenfell Cloth? The cloth itself, a dense weave of Egyptian cotton, can still be bought and sold. Yet, the mediated life of Grenfell Cloth, the ways through which we can come to apprehend it as a transmedia object of historical inquiry, concern, and construction, foreground a processual understanding of mediation *as* materialization that enables a two-way back and forth of coemergence between the material and immaterial planes, as scholars such as

Karen Barad have intimated, that is always already present in the lived and always fleeting moment.[4] I'm not arguing here for a qualitatively "better" or more "accurate" or more "truthful" form of historiography; rather, I'm arguing for forms of historical practice that recognize the performative mediation that certain historical artifacts demand in order to devise more, following Bergson, "intuitive" ways of knowing the world.[5] Much like commodity-chain analysis in anthropology, I seek to describe a transmedia story that emphasizes rather than elides the process of mediation itself all the way down the (historical) line. I offer my own description of this transmedia story as part of the Grenfell Mission narrative in order to highlight how archival work can be a generative form of knowledge production that recognizes its relationality with those archival objects and processes themselves. Thought of as a garment, I'm offering to turn the history of Grenfell inside out and let the marks of its making show in order to bring to light how narrative agency can be shared across a number of actors that can displace the subject-object dualism.

*

Beyond its embodied ontologies across archival and digital records, the *idea* for Grenfell Cloth took root in Grenfell's experiences in the First World War. Wilfred Thomason Grenfell was born in 1865 in the village of Parkgate in northwest England, and over the course of his life as a doctor, missionary, and enthusiastic proponent of a form of "muscular" Christianity—an ideal of the Victorian man of action, high minded, moral, purposeful, ready, and needing to "do"—he was genuine and, in guise of a sort of contextual sympathy, a genuine man of the late Victorian age. Among the aforementioned personas, Grenfell as author is shorthand for quite a prolific intervention into the world of the printed word and that of his cross-Atlantic and popular readerships. Over the course of his life, Grenfell wrote some two dozen books and four dozen journalistic pieces, with the majority of these being popular works that sought to shore up his vision of the venturing Christian, the man of action and principles, as well as the existence of a hearty stock of simple, good fishermen who were in need of altruism. In such narratives as *Vikings of Today, or, Life and Medical Work among the Fishermen of Labrador* (1895), *Adrift on an Icepan* (1909), and *The Romance of Labrador* (1934), Grenfell cultivated his own ability at survival in harsh conditions, the remote yet necessity-bound condition of Labrador, and the notion among multiple audiences of a desperate need for active cross-denominational philanthropy.

It was during Grenfell's medical studies at the London Hospital Medical College, which he entered in 1883, that he found his calling—or, to be truer to Grenfell's outlook, that he heeded a call. Under the tutelage of Frederick Treves, the London Hospital's senior surgeon, and to some extent Grenfell's ideal of a man in his athleticism and practical mindedness, Grenfell came to regard medical practice as a profession available to a broad field of endeavor rather than one confined to the private practice of the High Street doctor. In combination with a newfound religious idealism,[6] Grenfell was after a practical outlet for his faith. As he himself related across numerous publications over the course of his life, it was one night in East London when he was returning from consulting on a case in Shadwell that he happened, largely out of curiosity, to enter a tent meeting being led by the US evangelists Moody and Sankey. It was above all else D. L. Moody's perceived practicality that gave Grenfell a "determination either to make religion a real effort to do as I thought Christ would do in my place as a doctor, or frankly abandon it."[7] In 1892, he sailed for Newfoundland and Labrador and began to make his missionary mark on Great Britain's oldest colony.

In 1915, Grenfell attended the National Congress of Surgeons in Boston, and while there, he received an invitation to join the Harvard Surgical Unit, which was to sail from New York City in support of the Royal Army Medical Corps. By January 1916, Grenfell would be on the front. There, he observed that more damage was being done to the soldiers by exposure and the physical deterioration of trench life than by the enemy. Wet mud and low temperatures were producing trench foot, as well as a form of bacillus that could lead to gangrene. Grenfell's attention focused on the design of the uniforms themselves.[8] As he wrote in a letter from the front to his mother:

> I feel a little odd—the uniform is most uncomfortable. Very good to *look at*—but for weather, for wind & water, for warmth & for fighting just about as silly as could be devised. What on earth do I want with heavy hard boots—hard beastly gaiters—great heavy leather belt and shoulder straps. They would all freeze you in Labrador & would catch in everything that came along, & the cloth would soak mud & let water through & be cold & beastly.[9]

Grenfell was convinced that the conditions he had endured in Labrador, especially while at sea during his seasonal medical cruises along the coast, were just as harsh as those found in the trenches of the First World War. In an article published in the *British Medical Journal* of January 15, 1916, and entitled "Notes on Clothing Against Cold," Grenfell lays out the design, fabrication, and use of garments originally destined for the Labrador fishery. The central premise of his claim is that, as he writes elsewhere, "man is a centrally heated machine and all that is necessary to keep warm in a cold country is to keep in the heat."[10] The conditions of water, wind, cold, and heat are all environmental factors with which appropriate clothing can contend. While the clothing Grenfell sketches is of canvas and animal skin, both seal and deer (see figure 6), his wartime experiences prepared the way for the invention of Grenfell Cloth in the early 1920s. As with so many narratives of invention, military deployment and experience facilitated its emergence into wider sociocultural-civilian worlds.

This process of environmental mediation, of the extraction of selected experiential conditions met with in Labrador that would

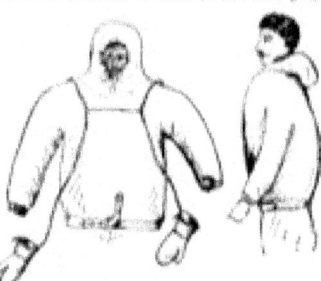
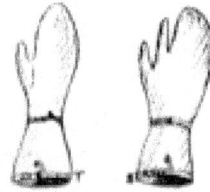
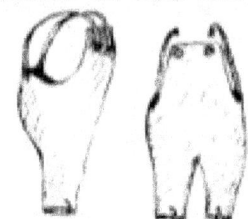

Figure 6. Wilfred Grenfell, "Notes on Clothing Against Cold," *British Medical Journal*, January 15, 1916.

then subsequently be reified in a commodity, had a precursor in the work of Clarence Birdseye. Today, these conditions are known under the brand Birds Eye vegetables, and their relationship to the Labrador winters lie in a distant if recast and mediated past.[11] Having secured the backing of a New York City investor (through Grenfell's personal intercession) to start up a fur farming venture, Birdseye spent his first winter on the Labrador coast at Muddy Bay from 1912 to 1913. He had bought the abandoned Révillon Frères fur trading post and went about building fox and mink pens. In his retelling, it was during that first winter that he began experiments in the quick-freezing of vegetables. "When I went to Labrador," Birdseye writes, "I knew nothing of the virtues of quick-freezing."[12] Birdseye's method took shape by observing "natives" catching fish in extreme temperatures. The fish would freeze "almost as soon as they were taken out of the water," and then months later, when they had thawed out, Birdseye observed that "those fish were so fresh that they were still alive!"[13] This temperature-related revelation set Birdseye to work on quick-freezing cabbages in Labrador. Upon returning to the United States, Birdseye would continue to develop and refine the naturally occurring technique that he had observed. He was after the re-creation of that moment of near-instantaneous freezing that could secure those qualities of flavor and texture across meat, fish, and vegetables. If in Labrador, Birdseye would postulate, through the instant of freezing produced by environmental climactic conditions, "time had literally stood still, . . . why couldn't similar conditions be produced by science?"[14] After ten years of development and the coordination of satellite infrastructures of refrigeration (for the transportation, storage, and display of Birdseye's frozen foods), and working on the cultural assumptions surrounding their freshness and nutritional properties, Birdseye would be able to proclaim the establishment of "a major new industry."[15] This bringing of Labrador's climactic realities to households across the United States was in some ways a much broader instance of environmental mediation than that produced by Grenfell Cloth. In Birdseye's estimation, quick-freezing "has been the most important development in the food industry since the invention of canning."[16] And while Birdseye's invention did not directly inform Grenfell's efforts to develop a Labrador-ready cloth, it was consistent with the doctor's attempts to use his missionary work as a field for the production of experimentation, with the majority of these efforts revolving around agricultural testing in order to determine the potential limits of the food supply in the harsh North Altanctic climate. Given that the majority of the prevalent diseases among the fisherfolk, such as beriberi, stemmed from malnutrition, the introduction of new cereals (tested at the US government's experimental agricultural station in Rampart, Alaska, and grown in latitude 63° 30 feet) and other crops (such as the soybean, which Grenfell, following research undertaken by the Ford Motor Company, saw as having multiple applications, most notably in the production of glass) were meant to diversify the local diet while thriving in its specific climate. Over the 1920s and 1930s, Grenfell would collaborate with agricultural researchers from the United States, Scotland, and New Zealand, among others, to experiment with various scientifically generated crops and soils that would see eventual application in northern Newfoundland and Labrador.

Grenfell Cloth proper emerged, so our story goes, in 1923. Part of Grenfell's work as superintendent of the mission entailed seasonal lecture tours of the United States, Canada, and the United Kingdom. During his illustrated lecture tour of 1922, Grenfell spoke at Burnley Hall in Lancashire. He had come to Lancashire at the invitation of Walter Haythornwaite, a local mill owner. In the early 1920s Anglo-Saxon world, Grenfell was an established celebrity quite nearly on par with Baden-Powell in the imperial imaginary. He was not an obscure evangelist working on an exotic population; rather, he was an active agent of acceptable and somewhat genteel if esoteric social reform for the colony's settler fisherfolk, the vast majority of whom emigrated from the west coast of

England and Ireland. At Burnley Hall, Grenfell spoke of the need for more appropriate clothing for the mission's own staff and the attendant foreign workers and volunteers that the mission asked to carry out that work in harsh seasonal conditions. Reliant on indigenous means of transportation such as the dogsled to reach remote communities in need of medical aid in winter, and without a ready supply of indigenous clothing nor the knowledge to make that clothing, the mission needed garments that could perform under demanding environmental conditions of which they had only a fleeting acquaintance, as opposed to the evolutive longevity of histories of indigenous environmental adaptation. After a year of experimentation, the Haythornwaite mill came up with Grenfell Cloth. As with many branding strategies, the *Grenfell* of Grenfell Cloth was meant to impart to the textile (and the coats, pants, and other articles of clothing that would stem from it) an authenticity of having passed the environmental "test" on the ground in Labrador. By putting on your Grenfell, you were, at least semantically and metonymically, draping yourself in the muscular Christianity embodied by the Grenfell Mission. And, of course, you were to be dry, warm, and immune to the elements at large.[17]

Haythornwaite and Grenfell, after some legal back and forth, agreed on terms that would have a small portion of every sale of an article of Grenfell Cloth go back to the mission. In addition to this, the Haythornwaite mill would provide the mission with cloth at cost, as the mission's industrial department, largely seeking to employ the women of coastal communities with wage labor, would actually produce a clothing line of their own for sale through their international philanthropic networks.

In the Grenfell Mission's industrial apparatus, the production of Grenfell Cloth outdoor garments directly on the coast came a distant second in scale to the production of hooked silk-stocking mats.[18] Mat hooking was an established local cultural practice through the settlers' Irish, Scottish, and English roots, but it was an early collaboration with Jesse Luther, an iconic figure in histories of occupational therapy and craft industrialism, at the turn of the century that would solidify and insitutionalize the production of mats for international sale.[19] A mission industrial worker would circulate throughout the coastal communities covered by the industrial department and distribute materials and hooking kits so that local women could enter into their system of production. The appearance of silk as a material for the mats came in the 1920s. This stock of materials was largely donated by women in metropolitan centers, particularly New York City. They would send their used silk stockings to St. Anthony for redistribution to the craft production center (locally known as "the Industrial") and its group of female mat producers. This distributed, domestic mode of production allowed women to create the mats in their seasonal downtime (limited as it was), and the mats eventually became one of the colony's best-known craft exports. "Our women in Labrador are always looking for work in the hard times of the winter," Grenfell writes, "and in order to help them while they maintain their perfect self-respect we have during the past years been giving them remunerative labor of this kind, and it is both a real help to our people and a real blessing to anyone who gets it."[20] Overall, the mission's industrial department, that Grenfell established in the 1890s, was essentially a home-craft industry that sought to combine the advantages of a readily available labor force (women at home, especially in the winter months; patients convalescing; and fisherpeople that had suffered a debilitating illness or injury that did not allow them to return to the fishery and had to find another type of employment) and locally available materials (in addition to donated materials, these included local supplies of wood and a stone known as labradorite). The department is one of the most enduring legacies of the mission, with handicrafts still a part of the local economy in northern Newfoundland.[21] Moreover, hooked mats, in the Grenfell Mission's publicity and fundraising campaigns of the 1920s and 1930s, figured as prominent, aestheticized examples of the work undertaken by this diverse (and deserving) labor force. The mats received their legitimation and a form of philanthropic sanction that Grenfell Cloth would also strive for and manufacture, by so clearly being the intricate handiwork of fisherfolk laborers using the donated materials of metropolitan concern.

For the case of Grenfell Cloth, a new and untested technology, Grenfell and Haythornwaite, in order to produce this sense of social legitimation, very astutely used testimonials from established explorers, public figures, and others to consolidate Grenfell Cloth's legitimacy as a weatherproof textile. One of the most notable instances of this process of legitimization came with Grenfell Cloth's official adoption as the garment of choice for the Royal Geographical Society's (RGS) Everest expeditions of 1933, 1936, and 1938. While these, of course, were not successful expeditions, they nonetheless went some way toward establishing the textile as a supposedly "advanced" and "authentic" technology that could ensure survival in the harshest conditions. In addition to the RGS's Everest expeditions, Rear-Admiral Byrd sought to have the cloth for his 1932 Antarctic expedition. Grenfell would try to assure him in a letter that "it certainly is, to-day, as water-proof as anything can be made."[22] To some degree, as a headline from the *Manchester Dispatch* of that same year attests, "Cotton to be Worn on Polar Dash: Admiral Byrd Orders Jumpers from Lancashire: The 'Grenfell Dickies,'"[23] the cloth also allowed for a reimagining of the limits of cotton as an extreme weather textile. As opposed to furs and treated hides produced by a given "nature," Grenfell cloth was "invented" by a local manufacturer and yet able to withstand the harshest "natural" climactic conditions. This expanded understanding of human agency and new materials is echoed by the *Manchester Evening News* of January 19, 1932: "Wonder Cloth Invented: Adapts Itself to Every Climate: Cotton Triumph: Dyeing Alone Took Seven Years."[24] Grenfell also solicited endorsements from the likes of explorers such as Henry Watkins. Through the influence of British alpinist Noel Odell, Grenfell sent Watkins a "scientific account" of the cloth's merits, with this last piece of documentation meant to sway the explorer away from using Burberry on his expeditions.[25] Moreover, Grenfell would emphasize the local economic benefits that could accrue to the people of Lancashire, since they were economically "on the rocks" and in need of national support.[26] Again, the popular press would seek to shore up a common societal vision of unlikely geographic connection, with the *Daily Mail* of January 20, 1933, proclaiming: "Looms Help for Climbers."[27] This bringing together of Lancashire and Everest, "Lancashire looms are helping to conquer the highest mountain in the world,"[28] was a typical imperial trope of scalar connection.[29] The suits made of Grenfell Cloth to be worn on the Everest expedition, weighing some two-and-a-half pounds, were designed by E. H. Taylor of the firm of Messrs. Baxter, Woodhouse, and Taylor, Ltd. However, they were "made," literally woven, by toiling Lancashire textile workers, whose hands were felt if not seen in such imperial endeavors.

These sorts of endorsements would also establish the missionary credibility of Grenfell Cloth and allow it to compete with rival textile manufacturers such as Burberry. Haythornwaite would also have new looms manufactured to accommodate the actual making of the cloth. As Grenfell notes, "They had to make special looms to force the amount of twist into the small amount of space and yet keep the cloth as smooth as can be."[30] This connection between "special looms" and the various end uses of the cloth led Grenfell to solicit the likes of Baden-Powell in an effort to convince him of Grenfell Cloth's utility for the Boy Scouts.[31] Drawing the parallels between its use in the Arctic, its important-if-experimental place in Byrd's Antarctic expedition, and the Prince of Wales' gifted golf jacket, Grenfell sought to bring the Scouts' clothing department into the orbit of imperial geographical extremes and royal sanction. Within the Grenfell Mission's operations, the cloth was an outcome of circulating within a margin of Empire and was an additional by-product of missionary action.

Over the course of the 1930s and 1940s, Grenfell Cloth would maintain its place as a reliable textile in the British clothing market. Its raincoats and sportswear marked the new social relations taking shape around labor practices, leisure time, and class-based cultural practices.[32] Grenfell Cloth could become "Clothing for All," as a Haythornwaite pamphlet states, and could bring an invocation of remarkable human feats—at their nexus with machine-enabled forms of conquest, including of lands, oceans, heights, speeds, and distances—as well as raincoats that were readily available to metropolitan consumers for metonymic association (see figure 7a). The Haythornwaite pamphlet reassembles the Grenfell Cloth story by mobilizing descriptions of the cloth's properties and direct citation from testimonials by Grenfell, Lady Grenfell, the Oxford University Ellesmere Land Exhibition, Robert North of the Explorers Club of New York, and, finally, anonymous endorsers. Moreover, the pamphlet also pairs up these human feats with firsthand testimonials from aviators, racing motorists, and explorers to inculcate a sense of all-pervasive confidence in the cloth's ability to withstand any and all uses and conditions. Coming under the belief that "a modern world uses Grenfell Cloth," Haythornwaite & Sons Ltd. would use the increasingly "modern" means of mass advertising to bolster sales and propagate stories of the cloth's far-flung use. In a circular letter from the 1930s, addressed to a "Dear Sir or Madam," Haythornwaite, following the firm's established formal narratives, cites Grenfell Cloth's use by Grenfell Mission workers, Byrd's expedition to the Antarctic, and expands it to include the North-West Mounted Police and representatives of the Hudson's Bay Company.[33] For Haythornwaite, "these things speak for themselves," with *things* designating a nebulous use-value enabled by such reliable institutional actors. In the 1940s, the firm also advertised in such print publications as *Punch*, *Vogue*, *Nursery World*, *Tatler*, *Parents*, the *British Medical Journal*, *Sketch*, and *Accountant*, while they also kept track of their relative success in each venue, with *Vogue* and *Nursery World* offering the best inquiry-to-purchase ratio (with an overall average of 45 percent of inquiries resulting in orders).[34]

Figure 7a. Promotional pamphlet for Grenfell Cloth, ca. 1930. Provincial Archives of Newfoundland and Labrador. *Source*: photo courtesy of the author.

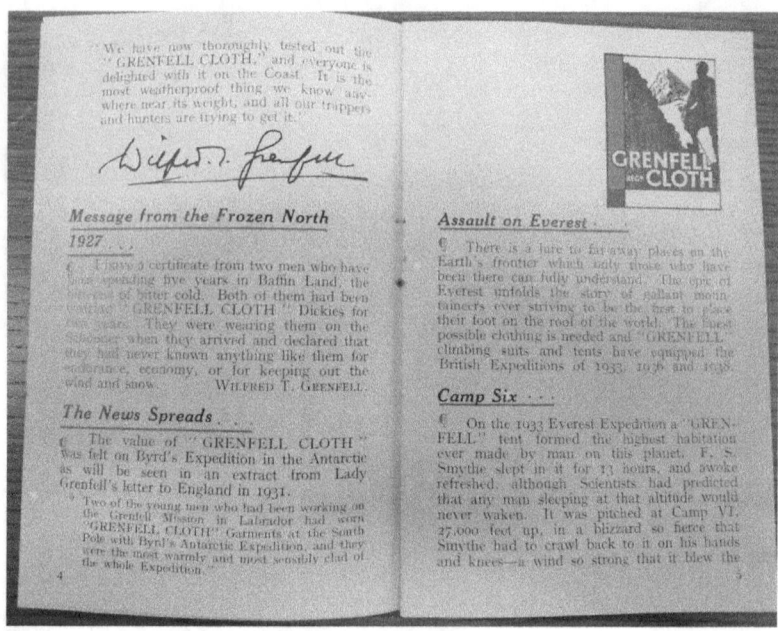

Figure 7b. Promotional material for Grenfell Cloth, ca. 1930. Provincial Archives of Newfoundland and Labrador. *Source*: photo courtesy of the author.

The Second World War made it difficult for the firm to both secure a ready supply of high-quality cotton for their manufacturing but also to ship their bulk orders of Grenfell Cloth to the mission's headquarters in Newfoundland. Requiring an "essential purpose" to receive the shipment of cloth, the IGA had to contend with shortages.[35] Eric Haythornwaite, Walter's son, having taken over the firm upon his father's death in 1942, would refuse to move production to either the United States or Canada:

> With the cloth made in our own mill everything is under our direct supervision, the type of cotton and the particular spinners selected together with all the processes in our own mill make it easier for us to maintain the quality upton [sic] the highest possible level, whereas making the cloth abroad would tend always to some deterioration which we have at all times wished to avoid. Grenfell represents the very best of everything in a piece of cloth.[36]

Like many forms of industry contending with and emerging from the Second World War, the Haythornwaite mill was holding onto a production model that would soon come up against a global marketplace for goods and for the manufacture of those goods themselves.

PERFORMING GRENFELL CLOTH, THE SLANT CUT

What does my narrativizing of the emergence of Grenfell Cloth ultimately point to? On the one hand, it marks the binding practice of discursivity itself in historiographical work. Yet, on the other hand, how can, or even should, we get around this discursive-representationalist bias when we *do* history? Are there ways that we can actually come to know, apprehend, and recognize if not the cloth itself then its contemporary manifestation as a phenomenon of mediated inquiry? Many of these questions go toward thinking about Grenfell Cloth in particular as a transmedia story, in King's understanding of the term, that can circulate in more performative analytical ways.

What makes of Grenfell Cloth a transmedia story are the ways in which it continued to materialize anew after the Second World War. While maintaining its place in the English clothing market in the decades after 1945, even garnering the sought-after mark of royal manufacturing assent, the Haythornwaite family mill in Lancashire would succumb to the pressures of the international textile market, finally being sold, both the mill and the patent of Grenfell Cloth itself, to a Japanese cashmere company in the early 1990s. The dominant, in being most accessible for us as readers and writers of history, story of the mill is, for the moment at least, preserved and commemorated in or at, if taken in more spatializing terms, one of those marginal sites of our digital worlds—a personal webpage (see figures 8a and 8b). Through rudimentary forms of digital representation, the Haythornwaite family lineage can reclaim its place in the textile's story. This representational strategy and artifact are making a media claim and taking the

Figure 8a. Screenshot of the home page of the Haythornthwaite family website, htttps://wwww.haythorwaite.com/index.html. *Source*: photo courtesy of the author.

Figure 8b. Screenshot of the "Grenfell Today" page of the Haythornwaite family website, htttps://www.haythorwaite.com/grenfelltoday.html. *Source*: photo courtesy of the author.

Grenfell Cloth story into its own world of semiotic, semantic, and distributed if embodied representational values (i.e., appearing on screens somewhere at some points in time). To reconstitute the Grenfell Cloth story, then, is to mine the cloth's most recent medium-specific incarnation, and, through that process, find discursive resonances that complicate its constitutive narrativity.

As a historical source, the internet as a medium offers a critical specificity in the forms of longevity, punctuality, and general sense of a lack of permanence and legibility that its modes of publication enable. "Like the serial media that it partially incorporates or 'remediates,'" as Gitelman claims, "the Web represents time and simultaneously produces temporalities for its users; it records and performs."[37] In this sense, the Haythornwaite family site is both a record and a performance of one incarnation of the artifact we are calling Grenfell Cloth. The site allows for one particular story (and representational mode) of the textile to take shape and emerge, to become a digital facticity in and of itself. If, as www.haythornwaite.com asserts, "this web site is run by a family of the name Haythornwaite,"[38] the site's pages become digital metonyms for the Haythornwaite name itself. The "History of the Name"[39] can be read along with a version of the Grenfell Cloth story—one that pivots around the period during which the brand was family-owned and produced.

For King, who in turn extrapolates out from Henry Jenkins' framing of "transmedia storytelling" as a form that "unfolds across multiple media platforms, with each new text making a distinctive and valuable contribution to the whole,"[40] transmedia storytelling is also a way of performing "queer transdisciplinarities."[41] This queering is not done on the level of identities; rather, for King, various kinds of academic practices are in the process of being remade and are producing transdisciplinarities that "require us to attend to, to learn to be affected by the political economies of knowledge worlds, to how interlinked the economies of entertainment, knowledge laborings, globally restructured academies, governmentalities, and the infrastructures of communication are now."[42] While the transmedia movement I trace is an *across* that looks at historiography as a process of mediation as materialization (as opposed to Jenkins' understanding of a transmedia story as a multiplatform entertainment-franchising effort), it is a form of historiographical practice that can point to the ways in which the relational dynamics between material-archival and digital-representational modes are being renegotiated. "Amid an unrelenting contextual historicism—the developmental arc or time of 'late' capitalism along with the unshakable, concomitant ideology of progress—the Web," as Gitelman contends, "helps to pose the question of history itself."[43]

To return to one of those complicating discursive resonances, Grenfell Cloth has been reappropriated and most recently deployed as a heritage-textile commodity by a company known as Grenfell England (see figures 9a and 9b). Within that nebulous world of digital intellectual property rights, Grenfell England recasts and literally recites the Haythornwaite story (and mission story, too, to some extent), as the texts on both websites are nearly *identical*, in order to position Grenfell Cloth as a uniquely and longstanding British commercial enterprise and experience:

Made in London:

While the exploratory spirit of Sir Wilfred Thomason Grenfell, the man after whom Grenfell Cloth was named, to one of [sic] most inhospitable outposts [sic] on the Labrador coast of Newfoundland [these are two separate landmasses], the current-day company is proud to represent the best in British manufacturing.

We produce quality clothing in the East of London, the textile and garment capital of the UK. From the moment we undertake a commission to the moment it reaches the end client, we demand the same exacting production and service standards observed since Grenfell Coats, Haythornwaite & Sons Ltd. was established.[44]

Figure 9a. Screenshot of the home page of the Grenfell England clothing company website, htttps://grenfell.com. *Source*: photo courtesy of the author.

Figure 9b. Screenshot of the "Made in London" page of the Grenfell England clothing company website, htttps://grenfell.com/pages/made-in-england. *Source*: photo courtesy of the author.

It is in this sense that Grenfell Cloth is a transmedia story that enables particular forms of mediated knowledge to emerge. It is not so much about the historian or other chronicler touching and seeing the cloth. Rather, it is about apprehending its status as a precarious and dynamic artifactual process that can travel from an inauthentic-imperial historicity of, in a way, environmental mediation (as it brought Labrador's climactic realities to wider networks of philanthropic and imperial concern), and how it came to travel on to become a primarily if not exclusively heritage commodity, made manifest through digital representational methods that are both competing and contestatory, as the Grenfell England and Haythornwaite representationalist strategies show. This is, in part, a result of the dominant "present tense of the World Wide Web," with its evolutive interpretive modes and modalities always already ongoing and on the move.[45] As an interpretive space, the internet is a volatile framework that is constantly morphing through system-user interactions. Yet, as a discursive space for historical facticity and historiographical practice, the internet is also a modality of the historical process itself, and, at that, one that is akin to communication as a phenomenon of mediated inquiry. As John Durham Peters notes, there are numerous commonalities between these phenomena known as "communication" and "history." Both are open questions that revolve around issues of interpretive distance, transmission across spaces and times, preservation, distortion, and recording, among a host of others.[46] As an object of evolving historiographical work, Grenfell Cloth is in and of itself an act of inscription, a record among others that is equally open to interpretation and triangulation:

> Recording is the act of inscribing something in enduring form; transmission is the act of sending a record across some kind of distance, whether space or time; and interpretation is the act of receiving transmitted records and putting them to work in the

present. Historical research is always a matter of triangulating record, transmission, and interpretation.[47]

How to undertake this triangulation when the object in question is a process that is always already materializing, when it is an evolutive production and performance that, by defying definition through intentional elision, also defies historiographical convention? One possibility might be to foreground how, following Durham Peters' recasting of Innis, "our knowledge of the past is a question of media."[48] In other words, historical work is also, and maybe foremost, a communication problem that has to contend with the "media" that are an integral part of its production, transmission, and reception. Those ever-problematic mediators—time and space—are always already marking the historical process as a process, smoothing its surfaces, and rendering its making if not invisible then conveniently subjective. For Innis, the "bias" of communication was not only a concatenation of forces working to disrupt genuine or "objective" communication—it was also the textile metaphor of a slant cut.[49] Historians, in that metaphorical world, have an obligation to interpret and create along the diagonal (see Darin Hayton's chapter in this volume for an analogously textured reading of the flinty semiosis Quakers imbued with their own ethics of care for mental patients in the early decades of the nineteenth century). To foreground the media at hand in interpretive practices, then, is to engage in the creation of a transmedia story. This is the simple recognition of agency in the making of hermeneutic worlds—that is, *we* is in fact more transparently an *I* that truncates, collages, and narrativizes. However, this is also the recognition that historical exegesis needs to "learn to be affected," as King puts it:

> New media infrastructures, boundary objects, and processes of learning also work across redistributed agencies, ones not located

simply in the consciousness of individual humans in seeming control, but rather ones emergent across materialities of social media old and new, together with beings and economies and knowledge workers and neurobiological systems, affecting and being affected.[50]

Apprehending processes of historical, political-economic, and representationalist formation such as Grenfell Cloth, in this view, requires a collaborative agency of interpretation. While the metaphoric possibilities of textiles are readily apparent and seem so ideally appropriate to new materialist readings of contemporary conjunctures, it is important to recall the peripheries of weaving or joining as reading-writing practices that engage with meaning-laden worlds in the process of their making. In this world, "the primary ontological units are not 'things' but phenomena—dynamic topological reconfigurings/entanglements/relationalities/(re)articulations," and, as a consequence "agency is not an attribute but the ongoing reconfigurings of the world."[51] The case of Grenfell Cloth in its performative instability asks of us how and why to intervene, as well as how descriptive apprehensions matter. What could be thought of as historiographical "textile media" such as Grenfell Cloth, in a transmedia story, are modes of apprehending our coconstituting worlds that can cut diagonally across narratives of closure—narratives of heritage commodities, family legacies, and missionary reforms. In a transmedia story, textile media can accomplish ways of reading that are ways of foregrounding a dynamic and tenuous discursivity that is always already giving way to the present reinvention of its own subject-object split. "Like any object of description," Peters writes, "the past is emergent."[52] This is a recognition of the performative dimensions of historiography. Grenfell Cloth, while a silky, strong, and weatherproof story to tell, is also an affected one in its ongoing redistribution of semiomaterial agency.

NOTES

1. International Grenfell Association Fonds, "The Labrador Ball: In Aid of The Grenfell Association of Great Britain and Ireland," Grenfell Association of Great Britain and Ireland, Provincial Archives of Newfoundland and Labrador, MG 63.323 Programs, Balls, Concerts, Matinees, 1934–1937.
2. Lisa Gitelman, *Always Already New: Media, History, and the Data of Culture* (Cambridge, MA: MIT Press, 2006), 5.
3. See Katie King, *Networked Reenactments: Stories Transdisciplinary Knowledges Tell* (Durham, NC: Duke University Press, 2011).
4. See Karen Barad, *Meeting the Universe Halfway: Quantum Physics and the Entanglement of Matter and Meaning* (Durham, NC: Duke University Press, 2007).
5. See Henri Bergson, *Creative Evolution*, trans. Arthur Mitchell (New York: Random House, 1941); Gilles Deleuze, *Bergsonism*, trans. Hugh Tomlinson and Barbara Habberjam (New York: Zone Books, 1988); Elizabeth Grosz, *The Nick of Time: Politics, Evolution and the Untimely* (Durham, NC: Duke University Press, 2004).
6. Wilfred Grenfell, *A Labrador Doctor: The Autobiography of Wilfred Thomason Grenfell, M.D. (Oxon.), C.M.G.* (Boston, MA, and New York: Houghton Mifflin, 1919; London: Hodder & Stoughton, 1920), 21.
7. Grenfell, *Labrador Doctor*, 26.
8. Ronald Rompkey, *Grenfell of Labrador: A Biography* (Toronto, Buffalo, NY, and London: University of Toronto Press, 1991), 185.
9. Rompkey, *Grenfell of Labrador*, 187.
10. Wilfred Grenfell, "'Grenfell Cloth' for Arctic and Tropic Wear," *Among the Deep Sea Fishers*, 26 no. 1 (April 1928): 34.
11. Birds Eye Corporation, "Bird's Eye Roots," accessed June 20, 2013, http://www.birdseye.com/birds-eye-view/history.
12. Clarence Birdseye, "The Birth of an Industry," *The Beaver*, September 1941, 12.
13. Birdseye, "Birth of an Industry," 12.
14. Birdseye, 13.
15. Birdseye.
16. Birdseye.
17. To some extent, Grenfell lived out an extemporized religious ideal that relied on these climate-derived qualities. What would become his best known book internationally, *Adrift on an Icepan*, published in 1909, recounts the story of Grenfell responding to a winter distress call from a dire case on a distant bay. Having attempted the more dangerous if shorter route across the unstable

bay ice to reach the patient in time, Grenfell got caught with five of his dogs on a pan of ice that was floating out to the open ocean. Grenfell managed to survive the more than twenty-four hours he spent on the drifting pan by killing three of the five dogs, fashioning their skins into a coat (he had rushed to the patient with so little forethought that he was still wearing his Richmond Rugby Football Club uniform under his outer clothing, the latter of which he had removed to gain better freedom of movement when he was struggling to escape from the unstable ice), and then using their bones to build a flag-pole of sorts to signal to potential rescuers on land. Within days, the story had been picked up by international wire services and became a "modern-day fable of endurance" (Rompkey, 148) that became the mission's most successful accidental publicity event in its history. For the canonical Canadian recounting of this episode, see Pierre Berton, "The Adventures of Wilfred Grenfell," *The Wild Frontier: More Tales from the Remarkable Past* (Toronto: McClelland and Stewart, 1978), 51–82. For more on this strain of imperial muscular Christianity, see J. A. Mangan, *The Games Ethic and Imperialism: Aspects of the Diffusion of an Ideal* (Harmondsworth, UK: Viking, 1986); and Mark Girouard, *The Return to Camelot: Chivalry and the English Gentleman* (New Haven, CT: Yale University Press, 1981).

18. See Paula Laverty, *Silk Stocking Mats: Hooked Mats of the Grenfell Mission* (Montreal: McGill-Queen's University Press, 2005).
19. See Ronald Rompkey, ed., *Jesse Luther at the Grenfell Mission* (Montreal: McGill-Queen's University Press, 1991).
20. Grenfell, "'Grenfell Cloth' for Arctic and Tropic Wear," 35.
21. See Grenfell Historic Properties, "Grenfell Handicrafts Online Store," Grenfell Interpretation Centre, accessed February 18, 2014, http://www.grenfell-properties.com/STORE/.
22. Letter from Grenfell to Rear-Admiral Byrd, April 8, 1932, International Grenfell Association Fonds, Grenfell Association of Great Britain and Ireland, Provincial Archives of Newfoundland and Labrador, MG 63.164, Grenfell Letters.
23. "Cotton to be Worn on Polar Dash: Admiral Byrd Orders Jumpers from Lancashire: The 'Grenfell Dickies,'" *Manchester Dispatch*, January 19, 1932.
24. "Wonder Cloth Invented: Adapts Itself to Every Climate: Cotton Triumph: Dyeing Alone Took Seven Years," *Manchester Evening News*, January 19, 1932.
25. Letter from Grenfell to Odell, April 8, 1932, International Grenfell Association Fonds, Grenfell Association of Great Britain and Ireland, Provincial Archives of Newfoundland and Labrador, MG 63.164, Grenfell Letters.
26. Letter from Grenfell to Odell.
27. "Looms Help for Climbers," *Daily Mail*, January 20, 1933.

28. "Looms Help for Climbers."
29. For comparative examples, see Jayeeta Sharma, *Empire's Garden: Assam and the Making of India* (Durham, NC: Duke University Press, 2011); Paul Gilroy, *Darker Than Blue: On the Moral Economies of Black Atlantic Culture* (Cambridge, MA: Harvard University Press, 2010); Ronald Hyam, *Understanding the British Empire* (Cambridge: Cambridge University Press, 2010).
30. Letter from Grenfell to Odell.
31. Letter from Grenfell to Baden-Powell, April 15, 1932, International Grenfell Association Fonds, Grenfell Association of Great Britain and Ireland, Provincial Archives of Newfoundland and Labrador, MG 63.164, Grenfell Letters.
32. See Raymond Williams, *Culture and Society, 1780–1950* (London: Chatto & Windus, 1958).
33. Letter from Haythornwaite & Sons Ltd. to "Dear Sir or Madam," n.d., International Grenfell Association Fonds, Grenfell Association of Great Britain and Ireland, Provincial Archives of Newfoundland and Labrador, MG 63.1980, Grenfell Cloth.
34. Letter from [sender cut off of page] to Haythornwaite, February 11, 1942, International Grenfell Association Fonds, Grenfell Association of Great Britain and Ireland, Provincial Archives of Newfoundland and Labrador, MG 63.1980, Grenfell Cloth.
35. Letter from E. Haythornwaite to Bremmer, Grenfell Association, London, January 8, 1942, International Grenfell Association Fonds, Grenfell Association of Great Britain and Ireland, Provincial Archives of Newfoundland and Labrador, MG 63.1980, Grenfell Cloth.
36. Letter from E. Haythornwaite to Spalding, April 14, 1944, International Grenfell Association Fonds, Grenfell Association of Great Britain and Ireland, Provincial Archives of Newfoundland and Labrador, MG 63.1980, Grenfell Cloth.
37. Gitelman, *Always Already*, 138.
38. Haythornwaite, "Home," Haythornwaite: A Long Name with a Long History, accessed June 20, 2013, http://www.haythornwaite.com (page no longer accessible).
39. Haythornwaite, "Home."
40. Cited in Katie King, "A Naturalcultural Collection of Affections: Transdisciplinary Stories of Transmedia Ecologies Learning," *Scholar & Feminist Online*, accessed June 20, 2013, http://sfonline.barnard.edu/feminist-media-theory/a-naturalcultural-collection-of-affections-transdisciplinary-stories-of-transmedia-ecologies-learning/4/.
41. King, "A Naturalcultural Collection of Affections."
42. King.

43. Gitelman, *Always Already*, 147.
44. Grenfell England, "Made in London," accessed June 20, 2013, http://grenfellengland.com/made-in-london?location=europe.
45. Gitelman, *Always Already*, 145.
46. John Durham Peters, "History as a Communication Problem," in *Explorations in Communication and History*, ed. Barbie Zelizer (London and New York: Routledge, 2008), 20.
47. Peters, "History as a Communication," 20.
48. Peters, 20.
49. Peters, 20.
50. King, "Naturalcultural Collection."
51. King, "Naturalcultural Collections."
52. Peters, "History as a Communication," 22.

AFTERWORD

OLD MATERIALS

Projit Bihari Mukharji

If you are reading this line, most likely you have already read the fascinating essays that precede it in this volume. You already know the multifaceted, complex, and situated work that the labeling of a material as "new" does. Therefore, in this brief afterword, I want to look at the other side of Jordan.

To call something "new" is of course to mark it off from that marked as "old," to deny their coevalness.[1] While claims about innovation and novelty invest the former label with particular forms of politics, social instrumentalities, and value, they also willy-nilly bequeath to its Other a set of negative possibilities. Further, by labeling certain materials nonmodern, archaic, old, and outdated, modernizers, perhaps unwittingly, open the door for these negatively framed materials to become orthogonally invested with a conservative code of traditionalism. What is "old" and "outdated" to the modernizer becomes the "authentic tradition" of the traditionalist.

These are not barren quibbles over terms. A spate of elections in recent times have brought to power a plethora of populist politicians who claim to speak for "authentic traditions," "old ways of life," and "how things used to be," and, at least in rhetoric, reject what is "new" or "modern." It is true that in action, as opposed to rhetoric, these figures do often embrace much that is seemingly new, such as military technology, multinational corporations, and the electronic media, but they tend to frequently do so by grandfathering these recent inventions as forms and items of "archaic modernity."[2] Thus, we suddenly find plastic surgery and space travel in the Vedas,[3] pineal glands and black holes in the Quran,[4] and the big bang theory in the bible.[5]

In an age defined by Donald Trump, Brexit, Jair Bolsonaro, Narendra Modi, Reccep Erdogan, Benjamin Netanyahu, and their likes, it is clear that the almost-monopolistic legitimacy that "newness" had enjoyed, at least in theory since the Enlightenment, has faltered. While innovation as a mantra keeps getting louder, and it would be utterly erroneous to write the epitaph of "newness" yet, it is also true that being "archaic" or "not-new" has jostled its way back onto the stage of history, culture, society, and politics. But what does any of this have to do with "materials"?

Many of the new political conflicts between the "new" and the "old" have found resonances in patterns of material usage. Some materials have been designated "new" and "modern," while other objects or substances have been opposed to them as "old" and "traditional." Indeed, it is through patterns of material use that the most divisive political conflicts of our times have become interpellated into the everyday lives of ordinary citizens.

One of the sites where this interplay of new and old, modern and archaic materials has played out most prominently, at least in South Asian history, has been around utensils used for cooking and eating. Over millennia, so far as archaeologists and historians can tell us, South Asian peoples have gone from using earthen vessels to terracotta, to a variety of early ceramics, to wares of manifold older metals and alloys such as silver and bell metal (*kansa*), to modern china or porcelain, to German silver and enamelware, to aluminum and steel, and eventually to plasticware. In the course of my own growing up in the region, I have experienced the shift from bell metal to china to steel and lately to plastic. Add to this a shift in utensils used for special ceremonial feasts, from the leaves of plants such as plantain or *shaal* (*Shorea robusta*) to Styrofoam. There has been enormous and rapid transformation of the materials in which we cook and eat our food in South Asia.

Yet, every time there has emerged a new material, there has also emerged a backlash against it. Interestingly, since commensality is one of the most productive sites for the inscription of social boundaries in the region, who cooks and eats in utensils made from what became a key marker of religious and caste identities and hierarchies. Thus, one observer writing in the 1930s reported that "Hindus, especially those belonging to the upper castes, will always use cooking utensils made of brass and other metals while Mohammadans [sic] generally use earthen pots. A Christian mostly uses aluminium and China clay pots."[6] These codes of commensality were particularly rigorously policed in upper-caste and elite households, where deviations were thought to cause ritual pollution (*apavitrata*).[7]

Naturally, the sustenance of this fractured social demand also ensured that the production of utensils in different materials continued side by side. Instead of the new materials pushing out the archaic ones, both continued to be produced and used by different segments of the population. The temporal split between new and old allowed a set of social instrumentalities, political agendas, and regimes of production to congeal around both labels respectively.

Like the issue of the consistency of new materials explored in the foregoing chapters, the old or archaic materials too had to be *made* consistent. Consistency was neither a given nor a stable state. It had to be achieved and kept up through repeated investments of time and labor. Materials distinguished by time, place, and methods of production were either lumped together or split up to create consistent categories of archaic materials. Most upper-caste Hindus throughout the large British Indian province of Bengal (mainly comprised of the entirety of the present-day Indian state of West Bengal and parts of Assam, along with all of Bangladesh) during the interwar years ate on utensils made from *kansa*. The method of production and actual constituents of these bell-metal utensils, however, had come to vary over the years. There were at least two broad processes by which the utensils were manufactured; namely, *dhala* (molding) and *pita* (hammering). These two processes depended upon distinctive ingredients. The kansa used in the production of molded utensils was considered a slightly inferior alloy whose technical name in the industry was *bharana*. Kansa proper was thus the name attached to the alloy used in hammering. It was a belief among the producers that the bharana alloy could not withstand the hammering, and that it was less durable. It was also thought to lack the luster of kansa proper and was hence touched up with artificial polish. Yet, even the so-called kansa proper was not always a uniform entity. It was widely held that the alloy varied with reference to the locality of production. The best kansa was thought to be produced at Khagra. Relatively inferior alloys, though still thought better than bharana, were produced in Bahrampur, Cuttack, and Dainhat.[8]

Over the decades, as demand had grown for kansa utensils, either through the accumulation of greater wealth amongst upper-caste families or the mimicry of their practices by lower-ranking castes, the social organization of production had also mutated. Like most other industries dating from the time before colonialism, kansa utensils were manufactured by a specific caste known as *kansaris*. The latter were thought to be a branch of the larger and older caste of *kamars* (ironsmiths). Yet, by the interwar years, the expansion of these businesses and the consequent opening of more workshops had led to a labor shortfall in the industry. This in turn had led to the employment of non-kansaris as apprentices.

Represented among these new recruits in the industry were such castes as the *kolhu* (oil presser), *bagdi* (soldiers and palanquin bearers), *kaibartas* (fishermen), and Muslims.[9] Of these individuals, several had also gone on to become master artisans themselves. Clearly then, the production, whose limitation to a particular caste was theoretically meant to ensure the ritual purity of these utensils, with rising demand had already slipped away from the kansaris. From the orthodox perspective, this might also have meant a variation in the "occulted materialities" of the kansa that embodied its ritual purity.[10]

These variations in ingredients, manufacturing processes, and the social identity of the manufacturers demonstrated that even in the case of this single well-known material (i.e., kansa), the problem of consistency could play out at multiple levels. Patently, this most authentically traditional material could still be manufactured by a number of distinctive technical and social pathways. That, notwithstanding all these variations, the final product was still recognized as traditional kansa was precisely because consistency was socially produced across these variations.[11]

The full range of actions through which such consistency was produced is beyond the scope of this chapter. But I will elaborate upon one particular way in which consistency was produced. I choose to offer this particular example of how consistency was produced because it also raises other issues about our own craft (i.e., history writing)—the craft that has brought this volume into being.

Historians, at least amateur ones, played a crucial role in building up the consistency of the "archaic material." They did this by both emphasizing the unity of the category, even when they saw variations, as well as by inserting the materials like kansa into a longer genealogy of materials used to make traditional vessels, just as chinaware or porcelain were sometimes inserted into a common category alongside earthenware and older ceramics.[12] Silver and bell metal were similarly, on occasion, clubbed together with aluminum and steel utensils by chroniclers. At other moments, and by other authors, these materials or sets of objects were distinguished and individualized.[13] By thus using historical narration as a way of organizing and reorganizing the categories to which particular materials belonged, authors were able to socially and discursively produce a level of internal consistency for the material itself. Kansa thus became a single, internally coherent material, by glossing over the variations and shifts in production processes, ingredients, and the social identities of the artisans who produced it.

Central to this production of consistency through historical narration was the way time was organized in these narratives. Irrespective of whether or not these historical works split or lump different materials, as historical works, they all posit a singular, chronological, and linear temporal frame organized around discrete points of beginning. Thus, even if some distinguished ceramic from the more recent chinaware, or bharana from kansa, the distinction itself is organized and rationalized around a linear timescale. The historian posits, as indeed I did above, that bharana is not akin to kansa because its methods of production and ingredients are distinctive and have a traceably distinctive genealogy. In so doing, however, I, like other historians, have also created new temporal points on the linear scale around which yet another set of arguments about "new" and "old" could be secreted. Yet another set of social instrumentalities, political imaginations, and regimes of production could be organized around these binaries. On the other hand, if another historian, in crafting a narrative of artefactual change, refused to countenance my distinction between bharana and kansa, they too would have to organize their unified category along a linear historical time. This would potentially defuse the polarization I had created but produce other points of difference between, say, kansa and silver or ceramic. The point I am trying to make, then, is that as long as we impose linear chronologies on materials, as indeed we need to in order to define them, there will also always repeatedly emerge a binary between the old and the new.

Instead of trying to find an impossible way out of this binary, I would suggest that we as historians accept our own complicity in the production of labels like "old" and "new." Such acceptance is not merely a matter of clearing our collective conscience but also an initial step toward being sensitive to the inescapable political possibilities bound up with historical scholarship. Defining a material as "new" or "old" both give rise to and resonate with particular social and political interests. We cannot write under the sign of some idealized notion of disinterested historical objectivity that is perfectly neutral or apolitical. We must recognize that whether we endorse the claims of newness for one material or subvert such claims for another, we are always doing so within a charged field of social, political, and economic interests.

This collection does much to unpack claims of newness for particular materials and interrogate the range of interests such claims serve; what I would like to emphasize, as an afterword to it, is that every claim of newness carries within it a distancing from something old as well. Old materials also have to be produced and reproduced as "archaic" or "old" materials. Meanwhile, oldness, too, serves its own constituencies of interest. As historians, we are not disinterested observers deciding between rival claims of old and new, but rather implicated within the force fields of rival interests.

In an age torn apart by violent rhetoric about innovation and tradition, historians have a moral duty to intervene and not just pretend to be disinterested accountants organizing life along a linear temporal scale. They must embrace the political possibilities, and pitfalls, of their profession and constantly interrupt the violence inflicted by the reification of labels like "old" and "new." They must insist that we see words like "innovation" and "tradition" not as self-evident homilies but as weapons with which social and political battles are fought. They must also accept that there are no final solutions in history: only a series of battles to be fought—against the banality of vacuous words like "old" and "new," "innovation" and "tradition," etc.

NOTES

1. Johannes Fabian, *Time and the Other: How Anthropology Makes Its Object* (New York: Columbia University Press, 2014).
2. Banu Subramaniam, "Archaic Modernities: Science, Secularism, and Religion in Modern India," *Social Text* 18, no. 3 (2000): 67–86.
3. Staff Reporter, "Ancient India Had Spacecraft Technology," *The Hindu*, December 16, 2000; Maseeh Rahman, "Indian Prime Minister Claims Genetic Science Existed in Ancient Times," *Guardian*, October 28, 2014, sec. World News.
4. "Quran Explains the Rules of Physics, Says Dr. Kayali," *Khaleej Times*, July 23, 2012.
5. Gerald Schroeder, *Genesis and the Big Bang Theory: The Discovery of Harmony between Modern Science and the Bible* (New York: Random House, 2011).
6. Shitla Prasad Saksena, "Cost of Living, Wages and the Standard of Living of Industrial Labour at Cawnpore," *Indian Journal of Economics* 17 (1937): 39–50.
7. Ravindra S. Khare, *The Hindu Hearth and Home* (Delhi: Vikas, 1976), 165.
8. Radhakamal Mukerjee, *The Foundations of Indian Economics* (London: Longmans, 1916), 231–35.
9. Mukerjee, *Foundations*, 235.
10. Projit Bihari Mukharji, "Occulted Materialities," *History and Technology* 34, no. 1 (January 2, 2018): 31–40.
11. Mukharji, "Occulted Materialities."
12. Phuldev Sahay Verma, *Mitti ke Bartan [Earthen Utensils]* (Prayag: Vishwaprakash, 1979); Ram Govindchand, *Pracheen Bharatiya Mitti Ke Bartan: Puratattva [Ancient Indian Earthen Utensils: Archaeology]* (Varanasi: Chowkhamba Vidya Bhavan, 1960).
13. Mukerjee, *Foundations*, 231–38.

Contributors

Darin Hayton is an associate professor of history of science at Haverford College, Haverford, Pennsylvania. He is also on the editorial board for Lever Press and deeply committed to the press's mission. Typically, his work focuses on the social and political uses of science technology in premodern Europe, and interrogates the ways that cultural values create science and how, in turn, that science shapes those values.

Scott Gabriel Knowles is professor and head of the Department of History at Drexel University, Philadelphia, Pennsylvania. His work focuses on the history of disaster worldwide. Knowles is the author of *The Disaster Experts: Mastering Risk in Modern America* (2011), editor of *Imagining Philadelphia: Edmund Bacon and the Future of the City* (2009), and he is coeditor (with Richardson Dilworth) of *Building Drexel: The University and Its City, 1891–2016* (2016) and (with Art Molella) of *World's Fairs in the Cold War: Science, Technology, and the Culture of Progress*.

Sharon Ku is an assistant professor in the Department of Engineering & Society at the University of Virginia. Her research and teaching focus on the geopolitics and sociology of standards/standardization. Her recent project takes on a comparative angle to examine the political and cultural roots of smart city standards in the United States, Taiwan, and China.

Projit Bihari Mukharji is an associate professor in the history and sociology of science department of the University of Pennsylvania. He studies the history of science in modern and early modern South Asia, with a particular interest in issues of subalternity. He is the author of *Nationalizing the Body: The Medical Market, Print and Daktari Medicine* (London, 2009) and *Doctoring Traditions: Ayurveda, Small Technologies and Braided Sciences* (Chicago, 2016).

Rafico Ruiz is the director of research at the Canadian Centre for Architecture in Montreal. He was recently a Social Sciences and Humanities Research Council of Canada Banting Postdoctoral Fellow in the Department of Sociology at the University of Alberta. His book, *Slow Disturbance: Infrastructural Mediation and the Promise of Extraction*, is forthcoming with Duke University Press.

Tiago Saraiva is an associate professor in the Department of History of Drexel University. He is the author of *Fascist Pigs: Technoscientific Organisms and the History of Fascism* (The MIT Press, 2016), and he is currently working on a collective manuscript called "Moving Crops and the Scales of History." He is the coeditor with Amy Slaton of the journal *History and Technology*.

Karen Senaga holds a doctoral degree in history from Mississippi State University and is an assistant professor of history and high school social studies at Pierce College in Lakewood, Washington. She is currently working on a manuscript tentatively entitled "Tasteless, Cheap, and Southern" on the farm-raised catfish industry in the United States.

Amy E. Slaton is a professor of history at Drexel University. She is the author of books on the history of building materials, standards, and labor relations, and on formulations of race in US engineering. She is currently writing a critical history of "diversity" in US science, technology, engineering, and math (STEM) education. She is a coeditor with Tiago Saraiva of the journal *History and Technology*.

José Torero is an engineer at University College London in the United Kingdom. He works in the fields of safety, environmental remediation, and sanitation, where he specializes in complex environments such as developing nations, complex urban settings, novel architectures, critical infrastructure, aircraft, and spacecraft. He worked on the World Trade Center collapse investigation, the Organization of American States human rights investigation into the murder of forty-three university students in Ayotzinapa, Mexico, and he is currently serving in the Grenfell Inquiry.

Patryk Wasiak is an associate professor at the Institute of History, Polish Academy of Sciences, Warsaw. He holds MA titles in sociology and art history and a PhD in cultural studies. He is also a former fellow of the Volkswagen Foundation, the Netherlands Institute of Advanced Study, and the Andrew W. Mellon Foundation. His research interests include the cultural history of the Cold War, history of home technologies, and industrial design.

Acknowledgments

This volume reaches completion after a long period of preparation, and I am indebted to all of the authors for their notable patience and profound engagements with their case studies throughout the book's many iterations. As the book took shape, conversations with Tiago Saraiva and Scott Knowles about the overall project inestimably expanded its historiographic reach, and the astonishing analytic provocations offered by Darin Hayton at every turn were invaluable. I thank Darin also for introducing me to Lever Press. Lever represents an unprecedented commitment to the publication of accessible scholarly materials, and I am deeply grateful to Beth Bouloukos and the press for their support of this volume as part of that commitment. The anonymous reviewers guided the book to far more incisive and authentically interdisciplinary content than I could possibly have arrived at otherwise. Finally, I thank Jesse Smith for his extraordinary editorial and conceptual contributions to the project from its inception. Without his participation and friendship, this collection would not have come to be.

Index

Note: Page numbers in *italics* indicate figures or diagrams.

2-methylisoborneal (MIB), 47, 49, 56, 61, 63, 64, 71n68

Acciavatti, Anthony, 56, 64
accuracy, 115n44. *See also* standardization
   of animal taste testers, 64
   defined, 115n42
   of nanoscale measurements, 76–77, 98
actinomycetes, 48, 49
Acton Ostry Architects, 162
Adorno, Theodor, 143
Adrift on an Icepan (Grenfell), 247, 267–268n17
aesthetics
   and design of technological artifacts, 197–198n7, 198n10
   in Italian manufacturing, 207, 214
   of Quaker asylum design, 178–179, 180–181, 186–187, 190–191, 192
   of wood construction, 151–152, 153, 156, 160
African Americans, association with catfish, 42, 65
agency
   of animals, 22
   of asylum patients, 178, 186, 195–196
   of construction materials, 179, 185
   of objects, 18
   and responsibility, 19
   and transmedia stories, 244–245, 265–266
Agricultural Cooperative Service (ACS), 55
agricultural research, 132–135, 251
air, and window design, 179, 184–185, 189–190
Alder, Ken, 21
algae, effects on catfish flavor, 47, 48
aluminum, 205, 215, 216, 273
Aluminum Dreams (Sheller), 215

American Society for Testing Materials, 59
American Wood Council, 156
Ammerman, G. R., 52–54
Anderson, Benedict, 207
Andrews, L. S., 55
animals, 22, 64. *See also* catfish
Appadurai, Arjun, 211, 221
architectural choices, 155, 199n12. *See also* asylum design, Quaker; tall timber construction
archival work, and Grenfell Cloth, 28–29, 240, 246, 262
Archives, Provincial, of Newfoundland and Labrador, 237, *238, 241, 242, 257, 258*
artisanal production, bicycle manufacturing as, 212–214, 227–228, 229–230
Arup (fire protection engineering firm), 166
asylum design, Quaker, 26–27, 177–196
  door locks, 191–195
  and Quaker understanding of madness, 177–181
  Retreat at York as model for, 181–184
  window sashes, 184–191
A-Team (sourcing and product-promotion group), 224
athletes, performance of, 204, 205–206, 209–210, 215
atomic clocks, 74, 114n31
atomic precision, 76–77, 114n31
ATR (CFRP material supplier), 224
Austria, timber production in, 158

Bailey, Liberty Hyde, 122, 132–134

Barad, Karen, 17, 18, 19, 246
Barker, Anna, 84
Barthes, Roland, 210–211
Beckert, Jens, 5
Bennett, Jane, 18
bharana (metal alloy), 274
Bianchi bicycle company, 223–224, 226
bicycle frames, 27, 203–230
  advantages of CFRP for, 205–206
  and country-of-origin stereotyping, 206–207, 208–209, 220–222
  and emergence of competitive cycling, 209–211
  evolution into lifestyle object, 223
  manufacturing of as artisanal production, 212–214, 227–228, 229–230
  and national identity, 204–205, 207–209
  Taiwanese manufacturing of, 218–220
Bicycling Science (Wilson), 212
Bijker, Wiebe E., 211
Bikeradar website, 220
Binderholz GmbH, 155
Birdseye, Clarence, 250–251
bland flavor of farm-raised catfish, 41, 43, 45, 54, 65–66
blogs and webforums. *See* websites
blue-mold fungus, 123–124
bodies, human, 56, 125–126, 182–183
bodies, of catfish, 47–48, 62
bolts. *See* door locks
boundary objects, 81–82
bridges, 154, 198n10
British Aerospace (defense contractor), 216
British Columbia, Canada, tall timber buildings in, 160, 162–163

286 INDEX

British Medical Journal, 248, 249
Brock Commons, Vancouver, Canada, 160, 162–163
Brooklyn Bridge, 154
budding of citrus seedlings, 136, 137–140
building codes and regulations, 159–163
   differing models of, 162
   fire design methodology, lack of, 163
   history of, 152–153
   resistance to changes in, 161
   subjectivity of, 154, 161–162
   and tall timber construction, 154, 159, 162, 163
   in various countries, 159
Building Committee for Friends' Asylum, 177, 187
building materials. *See* construction materials; natural building materials
bureaucracy
   and citrus growers' cooperatives, 138–140
   in Quaker asylum design, 180
bureaucracy, and development of nanosize standard, 73–111
   and bureaucratic indifference, 97–100
   and bureaucratic mundaneness, 103–107
   and bureaucratic rigidity, 100–103
   and characteristics of standardized scientific objects, 73–75
   and interagency collaboration, 83–90
   managerial role in, 94–97
   in NIST, 90–94

   and precision in nanotechnology, 75–79
   as series of actions, 107–110
   and standardization, 79–80
   and techno-bureaucratic objects, 81–83, 97–107
Burrows, Mike, 219
Byrd, Richard E., 254

California Fruit Growers Exchange, 122
   annual meetings of, 129–130
   bureaucratic structure of, 127–128
   commercial success of, 130–131
   cooperative formation of, 127–129
   department of bud selection, 139–140
   local associations for funding of packinghouses, 126–127
   Powell as general manager of, 131–132
Campagnolo, Tullio, 213
Campagnolo bicycle company, 213
Canada. *See also* Newfoundland and Labrador, Canada
   timber construction in, 158, 160, 162–163
capitalist economies/industrial capitalism, 4–5, 122, 127–129. *See also* labor relations
Carbon-Fiber Bicycle Development Project, 219
carbon-fiber-reinforced polymer (CFRP), 203–230
   advantages of, 205–206, 216, 222
   applications of, 216–217
   development of, 205, 216
   European manufacturers' use of, 223–225

carbon-fiber-reinforced polymer (*continued*)
  limitations of, 216, 222
  and Taiwanese national identity/COO stereotyping, 222–223
  Taiwanese production of, 217–220
carbon footprints, 157–158, 160
caretaking, Quaker ethos of, 178–179, 181–183, 195–196
castes, in India, 273, 274–275
categorization/taxonomic activities, 3–4, 7–8, 14–15. *See also* consistency; standardization
  of catfish flavor, 40–41, 57–64
  of experts and expertise within NIST, 95
catfish, 14, 21–22
  behaviors of, 22, 47–48
  bodies of, 47–48, 62
  feed for, 44–45
  subjective sensory experiences of consumers, 41–42
  wild flavors of, 46–47
catfish, management of flavor in commercial production of, 12, 39–66
  and catfish behaviors, 22, 47–48
  challenges of, 44–52
  and changes in industry, 44–46
  farmers' role in, 46, 50, 51, 52–53
  and off-flavors, 39–41, 45–46, 48–49
  and pond environments, 44–45, 47, 48, 51
  precision in flavor assessment, 57–64
  problem definition in, 52–57
  processors' role in, 42–43, 50, 51, 52–55, 66
  researchers' role in, 43, 48–50, 51–52, 57–58, 59, 66
Catfish Row (Vicksburg, Mississippi), 42
ceramics, 272–273
Cho, Dung-Song, 220–221
Chu, Wan-wen, 218
Citrograph (citrus industry newsletter), 130, 141
Citrus Experiment Station of the University of California (UC) in Riverside, 122–123, 134–135
citrus growers' cooperatives, 12, 15, 23–24, 119–144
  and agricultural survey as community-building tool, 131–134
  and blue-mold fungus, 123–126
  and budding variations, 137–140
  California Fruit Growers Exchange organization, 126–131
  and Dewey's work on communitarian experimentalism, 120–121
  labor system and damage of fruit, 126
  and landscapes, 119–120, 121
  performance records kept by, 138–140
  and pickers/laborers, 139
  and racial divisions, 121, 123, 141–144
  and rootstocks, 134–137
civic well-being, 23
class
  and flavors of farmed catfish, 41, 44, 46, 65–66
  and Grenfell Cloth, 252–253, 255–256
  and orange pickers, 126, 142–143

classification. *See* categorization/
taxonomic activities
climate change, 157–158, 170
  IPCC, 155, 157
cloning, 121, 136–137, 139–140
CLT (cross-laminated timber)
  advantages of, 156, 165
  and competing industries, 161
  defined, 156
  encapsulation of, 163
  fabrication of, 155–157, 165
  fire resistance of, 153, 158
  ICC approval of, 159
  manufacturer claims, 158–159, 163
  post-fire performance, 163
Cohn, David, 42
cold, 29, 250
  and clothing design, 248, *249*
  and window design, 181, 182, 185
Collins, Harry, 100–101
Colnago, Ernesto, 213, 224–225
Colnago bicycle company, 213, 214, 217, 223–225
Columbus (metal-alloy tubing manufacturer), 212
Committee on Fastenings (of Building Committee of Friends' Asylum), 192–195
communication. *See also* websites
  of flavor descriptors, 59–61
  and history, 264–265
  influence of, 237
  and interagency collaboration, 83–90
complexity
  of enforcing building codes, 161–162
  of engineered wood products in construction, 167
  of factors contributing to building material acceptance, 170
  of fire safety on timber construction sites, 169
  of interactions between construction materials and hazards, 163, 166, 168
concrete as construction material, 16
  advantages of, 152, 161
  as core for timber structure, 162
  environmental cost of manufacturing, 157
  historical dominance of, 160
  safety demonstration tests, 167
  safety questioned, 155, 161
concrete industry, 161
conditional prophecy, 10
consistency, 7–8, 9, 20. *See also* standardization
  and farm-raised catfish, 45–46, 62
  of old/archaic materials, 273–276
construction, modern, history of, 152–155
construction materials. *See also* asylum design, Quaker; concrete as construction material; wood as construction material
  engineered wood products, 156–157, 167–168
  history of, 171
  perceptions of desirability, factors affecting, 155
  social instrumentalities of, 16
  steel as, 152, 155, 157, 160–161, 165
  trends, major shift in, 160–161
  values, embodiment of, 167
construction sites, fire hazard in, 12, 168–169

INDEX   289

cooperative finance, 243. *See also* citrus growers' cooperatives
Coppin, Dawn, 22
Cornell College of Agriculture, 132, 133, 134
Country Life Commission, 122, 131–132, 133
country-of-origin (COO)
   stereotyping, 208–209
   and internet forums, 225–227
   and luxury goods, 214
   and "Made in Taiwan" products, 27, 218, 220–222, 223–226, 229
cross-laminated timber. *See* CLT
cultural identity. *See* national identity
customer relationship, between NIST and NCI, 92
cycling industry. *See* bicycle frames

Daily Mail, 255
Daston, Lorraine, 28
Dauncey, Hugh, 210
delamination and fire risk, 168
democratic cooperation
   in citrus growers' cooperative, 128–129
   and Latour/Dewey, 120–121
   and racial divisions, 142–143
   and standardization of horticulture, 132–134
Denis, Jerome, 6
descriptive analysis, 59–60
Dewey, John, 17, 120–121, 132–133
Dialectic of Enlightenment (Adorno and Horkheimer), 143
digital representation. *See* websites
documents, and scientific standardization, 82–83
dogs, as taste testers, 64

door locks, 178, 183, 191–195
Dreicer, Gregory K., 153–154

Edgerton, David, 4, 7
Edgewater, New Jersey, 169
Edinburgh Review, 181–182
Egerton, John, 44
Eiffel Tower moments, 153, 160, 166
electronic sensory devices, 62–64
Ellis, Merle, 65
embodied energy, of building materials, 157
encapsulation of flammable building materials, 158, 163
energy consumption of buildings. *See* carbon footprints
engineered wood products, 156–157, 160, 163, 167–168. *See also* CLT
Epperson, Bruce D., 212
European bicycle makers, 204, 206–207, 213–214, 223–228
Evans, Jonathan, 177, 188
Everest expeditions, 242, 254–255
Expert on the Spreadsheet (NIST Excel file), 95–97, *96*

fabric. *See* Grenfell Cloth
farmers and farmworkers
   laborers employed by orange growers, 125–126, 127, 139, 143
   role in management of flavor of catfish, 46, 50, 51, 52–53
FDA–NCI– NIST Memorandum of Understanding (2006), 83–84, *85*
Federally Funded Research and Development Center (FFRDC), 89, 113n27
Feynman, Richard, 75, 110
firefighter safety, 168–169

fire insurance, 169, 170
fire protection engineering, 165–166
fire risk in wood construction, 152–153, 157, 162, 163–164, 168–169
fire safety demonstration tests, 166–167
fish. *See* catfish, management of flavor in commercial production of
flavor. *See* catfish, management of flavor in commercial production of
Food and Drug Administration (FDA), 83–84, *85,* 92, 97
Formosa Plastics, 218
France, bicycle manufacturing in, 210–211, 213–214
Friends' Asylum (Philadelphia), 177–196
  door locks at, 191–195
  and Quaker understanding of madness, 177–181
  Retreat at York as model for, 181–184, 189
  window sashes at, 184–191
frozen foods, development of, 250–251
future, industrial capitalism's orientation toward, 5

Gaines, Richard, 193
gas chromatography, 49
gender, 4, 5, 20
geosmin, 47, 49, 56, *61,* 63, 64, 71n68
Gerber, Nancy N., 49
Giant bicycle company, 217, 218–220, 221–222, 224
Giro d'Italia, 210, 211
Gitelman, Lisa, 244, 261, 262
glass, in windows, 185, 186, 189
Glover, Joe, 39, 54

glulam (glued laminated timber), 156, 163
gold nanoparticle reference material (gold RM), 14, 22–23, 73–111
  as boundary object, 82
  importance of bureaucracy in standardization of, 77–79
  interagency collaboration on, 83–90
  and multiple-size measurement, 97–100
  and precision, 100–103
  release of, 107
  "Report of Investigation," 104–106
  robustness of, 103–107
  use in nanomedicine, 77
  users of, 104–107
Goodwin, Samuel, 193–194
government incentives, for wood construction industry, 158, 161
government-owned, contractor operated (GOCO) facilities, 84, 89, 112–113n13, 113n27
Green, Michael, 151–153, 160, 166, 170
Green Building Council, United States, 155
green building materials. *See* natural building materials
Grenfell, Wilfred Thomason, 246–248
  and agricultural research, 251
  and development of Grenfell Cloth, 244–245, 248, 251–252
  and endorsements of Grenfell Cloth, 254–255
  missionary work of, 243, 246, 247, 251–252, 253
  writings by, 246–247, 248, *249,* 267–268n17
Grenfell Association of Great Britain and Ireland, 240

Grenfell Cloth, 28–29, 237–266
    advertising for, 240–242, 256, 257–258
    development of, 244–245, 247–248, 251–252
    endorsements of, 254–256
    garment from, *239*
    and Grenfell England, 262–264
    sale to Japanese company, 259
    samples of, *238*
    story on Haythornthwaite family website, 259–261
    and transmedia stories, 28–29, 245–246, 259–264, 265–266
    use on Mt. Everest, *242*, *243*, 254–255
Grenfell England (textile company), 262–*263*
Grenfell Mission, 251–254
Griesemer, James, 81–82
Griffitts, Samuel P., 177, 187–188, 193
Grimm, Casey, 63
Grodner, R. M., 55
group identities, 17. *See also* national identity, and bicycle manufacturing
gypsum wallboard, 163

Haake, Steve J., 206
Hagley Museum and Library workshop (2011), 8
Hare, Geoff, 210
Hasslacher Norica Timber, 155
Haythornthwaite, Eric, 258
Haythornthwaite, Walter, 251, 252
Haythornthwaite & Sons Ltd. (textile manufacturer), 252, 255, 256, 258–259, 262
Hayton, Darin, 26–27

heating, ventilation, and air-conditioning (HVAC), 26–27
heavy timber construction, 164. *See also* tall timber construction
high-rise architecture. *See* tall timber construction
Hight Walker, Angela, 86, 88, 89
historians
    and "bias" of communication, 264–266
    investigation of innovation by, 4–5, 9–10
    production of consistency by, 275–277
historiography, of technology, 7, 14, 29, 134, 240, 244, 245–246, 262, 264–265, 266
history of construction, 153, 160–161, 162
homogeneity. *See also* consistency; standardization
    of orchards, 138–139
    in production of gold RM, 102
Horkheimer, Max, 143
horticulture, and democracy, 132–134
Hudson Bay Company (HBC), 240

IBC (International Building Code), 159, 163
ICC (International Code Council), 159, 161
Imagined Futures (Beckert), 5
imperfections, in natural building materials, 157
industrialization
    and cycling industry, 206, 209–215
    and historical values, 11
    and investigation of new materials, 2

Industrial Technology Research Institute (ITRI), 219
Inland Marine Underwriters Association, 163
Innis, Harold, 237, 265
innovation-centric historical literatures, 3–4
insanity, historical understandings of, 196–197n2
   and limitations of medicine, 181–182
   Quaker understanding of, 177–181
   and York Retreat, 181–184
insurers, of buildings, 169, 170
intercultural flow, 221
International Building Code (IBC), 159, 163
International Code Council (ICC), 159, 161
International Grenfell Association, 240, 243, 258
International Panel on Climate Change (IPCC), 155, 157
internet. *See* websites
IPCC (International Panel on Climate Change), 155, 157
iron, 154, 179, 183, 184, 185
Italy, bicycle manufacturing in, 207, 210, 211, 213–214, 233n28

Japan, building codes in, 159
Jeffery, William, 84
Jenkins, Henry, 261, 262
Johnsen, Peter, 59, 60–61, 62
Jones, Benjamin, 191

Kaiser, Debby, 91–92, 93–95, 97–99, 101
kansa (bell metal), 273, 274–276

Kennedy, John, 194, 195
King, Katie, 245, 261, 265–266
KLH Massivholz GmbH, 155
Knowles, Scott, 12, 24–25
Ku, Sharon, 21, 22–23

laboratories, material culture of, 73–74
laborers employed by orange growers, 125–126, 127, 139, 143
labor relations, 11–12, 15
Labrador Ball (1936), 240
laminated veneer lumber (LVL), 156
Lancaster locks, 193, 194
land-grant researchers, 50, 64, 134
landscapes, 119–120, 121, 130, 141
Latour, Bruno, 30, 79, 82, 112n6, 112n7, 119–120
Law, John, 16
Lechevalier, H. A., 49
LEED (Leadership in Energy and Environmental Design) building certification, 155
Leger, L., 48–49
light, and window design, 179, 184–185, 189–190
lightness of materials, 206, 215, 222, 237
lightweight timber construction, 156, 158, 164
Lloyd, Steven, 63
locks. *See* door locks
Look bicycle company, 217
Lorenzetti, Ambrogio, 119–120
Los Angeles, California, 169
Lovell, Richard "Tom," 50–51, 52–54, 56, 57–58
Lovely Bicycle! blog, 226
Lowell, Fran, 52

Lower, Abraham, 191
LVL (laminated veneer lumber), 156
Lynch, Michael, 81

MacKenzie, Donald, 101, 115n44
"Made in Taiwan" products, 27, 218, 220–222, 223–226, 229
madness, as historical understanding. *See* insanity
Manchester Dispatch, 254
Manchester Evening News, 254
manufactured and natural materials, intermingling of, 153
Marketing Research Corporation of America, 46
Marshall, Stanley, 55
mass timber. *See* engineered wood products
mass timber construction. *See* tall timber construction
materials, 8–14
    detection of, 16–17
    embodying values, 167
    forms of engagement with, 8–9, 19–20
    histories of, 154
Mathews, John A., 220–221
mat hooking, 252–254
Mavic bicycle company, 215
Mayr-Melnhof Karton AG, 155
McQuaid, Pat, 222–223
measurement, 22–23
    of catfish flavors, 56, 57–64
    of nanoparticles, 76–77, 83–84
medicine
    and architectural design, 199n12
    Grenfell's studies in, 247
    nanomedicine, 21, 78, 83–84
    and new understandings of insanity, 181–182
    and physics, 86–87
metals
    aluminum, 205, 215, 216, 273
    bharana (metal alloy), 274
    in bicycle manufacturing, 205–207, 209–215
    iron, 154, 179, 183, 184, 185
    kansa (bell metal), 273, 274–276
    steel as construction material, 152, 155, 157, 160–161, 165
metrology. *See* gold nanoparticle reference material (gold RM)
MIB (2-methylisoborneal), 47, 49, 56, 61, 63, 64, 71n68
microwaves, 54
Miller, Adrian, 42, 43
Mills, Hannah, 181
Minneapolis, Minnesota, 151, 160
missionaries, 12, 28, 243, 247, 251–254
Mission Inn, Riverside, California, 131
Mjosa Tower, Brumunddal, Norway, 160
Mody, Cyrus, 10, 81
Moody, D. L., 247
moral treatment (of the insane), 178–179, 181, 195–196
"muddy" flavor of catfish, 41, 42–43, 45–46, 57, 65–66, 67n5
Mukharji, Projit, 29–30, 120
multiple-size measurement of nanoparticles, 97–100

nail-laminated timber (NLT), 156, 160
nanomedicine, 21, 78, 83–84
nanoparticles, size measurement of. *See* gold nanoparticle reference material (gold RM)
nanotechnology, 1, 75–79. *See also*

gold nanoparticle reference
material (gold RM)
Nanotechnology Characterization
Laboratory (NCL)
and collaboration with NIST
on size measurement of
nanoparticles, 87–88, 91, 94–96
establishment of, 84
identity of, 89–90
operation of, 113n13
National Cancer Institute (NCI), 76
and 2006 MOU, 83–84
collaboration with NIST on size
measurement of nanoparticles,
91–92
expectations of nanoparticle size
standardization, 84–85, 86–87
operation of, 113n13, 113n27
organizational structure of, 88
National Fire Protection Association
(NFPA), 153, 162, 172n16
national identity, and bicycle
manufacturing, 12, 203–230
and Colnago bikes, 223–225
and country-of-origin stereotyping,
208–209
and European artisanal production,
213–214, 227–228
and industrialization in Europe,
207–208
on internet forums, 225–227
origins of materials, 206–207
and technological innovation, 207–
209, 210–211, 227
National Institute of Standards and
Technology (NIST), 77
and 2006 MOU, 83–84
collaboration on size measurement
of nanoparticles, 85–87, 91–94

founding of, 112n11
measurement service division
(MSD), 103–107
and multiple-value standard for size
measurement of nanoparticles,
97–100
nanotechnology standard working
group, 95–97
Nobel Prize winners from, 114n31
organizational structure of, 88–89,
92–94
and precision, 101, 102–103
statistical engineering division
(SED), 101, 102–103
National Nanotechnology Initiative
(NNI), 75–76, 83
National Stone, Sand, and Gravel
Association, 161
natural and manufactured materials,
intermingling of, 153
natural building materials
desirability in construction, 151–155,
157, 158
imperfections of, 157
subjectivity of, 167, 170
nature, 12–13
landscapes, 119–120, 121, 130, 141
and promotion of wood-framed
architecture, 151–152, 153, 155
Nature-Study Idea, The (Bailey), 132
negative carbon footprint, 158
Newfoundland and Labrador, Canada
agricultural experimentation in,
251
development of quick-freezing in,
250
and Grenfell Cloth, 28, 243
Grenfell's missionary work in, 247,
248, 251–253

Newfoundland and Labrador, Canada (*continued*)
   Provincial Archives of, 237, *238*, 241, 242, 257, 258
new materials, understandings of, 2–5, 19–20
newness
   social instrumentalities of, 9, 271
   value of in construction materials, 154, 171
news media, architectural, 160
NFPA (National Fire Protection Association), 153, 162, 172n16
NLT (nail-laminated timber), 156, 160
Norway, timber production in, 158
"Notes on Clothing Against Cold" (Grenfell), 248, *249*
novelty, 3, 8–9, 10, 13, 25, 29–30, 76, 166, 271

Objects and Materials: A Routledge Companion (Harvey et al.), 11, 17
off-flavors (of catfish), 40, 67n5
old materials, 271–277
   wood as, 153, 154, 170
oranges, 12, 23–24. *See also* citrus growers' cooperatives
   blue-mold fungus effect on, 123–124
   handling of, 124–126
   rootstocks of, 135–137
   strains of, 137–138
organic building materials. *See* natural building materials
original equipment manufacturers (OEMs), 207. *See also* country-of-origin stereotyping

packinghouses (for oranges), 126, 127
parliament of things, 119–120
Patri, Anil, 86–87
Paul, Joseph M., 177, 187–188, 193
pendulum locks, 193–194, 201n49
Perez, Karni, 45
performance-based model, of building codes, 162
Peters, John Durham, 264–266
Peugeot, Jean Pequignot, 213
Peugeot bicycle company, 213
pharmaceuticals, 14
physics, 86–87, 114n31
Pickering, Andrew, 16
plastic, 154, 216, 218, 273
polyculture, 51–52
polystyrene, 87
Pontille, David, 6
populist movements, 128, 272
Porter, Theodore, 79, 80, 112n6
Postel, Charles, 128
post-fire building performance, 163
Pottage, Alain, 7–8, 21
Powell, G. Harold, 123, 124, 125, 126, 127, 131, 132
Practical Hints (Tuke), 185–186, 192
precision, 75–79, 100–103, 110, 115n42
precompletion risks, in construction, 169
prescriptive model, of building codes, 162, 172n16
Progressive Movement, 131–132

Quaker Oats Company, 39–40
Quakers. *See* asylum design, Quaker
quick-freezing, development of, 250–251

racial difference, 24
   and citrus industry, 121, 123, 141–144

and flavors of farmed catfish, 41, 42–43, 46, 65–66
racing bicycles. *See* bicycle frames
raw materials vs finished products, 14
Ready Mix Concrete Association, 161
Reconstruction in Philosophy (Dewey), 133
regulations, building. *See* building codes
Reid, Carlton, 228
religion
    and commensality, 273, 274
    missionaries, 12, 28, 243, 247, 251–254
reproducibility. *See* consistency; standardization
Reynolds Cycle Technology (metal-alloy tubing manufacturer), 212
River Beech Tower, Chicago, USA, 160
Road Bike Review website, 225, 226
Roberts, Lissa, 4, 15
Roebling, John Augustus, 154
Roosevelt, Theodore, 122
rootstocks, 134–137
Royal Geographical Society (RGS), 254
Royal National Mission to Deep Sea Fishermen (RNMDSF), 243
rubber, 16
Ruiz, Rafico, 12, 28
Russell, Andrew, 10

safety
    of concrete as construction material, 155, 161, 167
    fire safety, 166–167, 168–169
    perception of safety for high-rises, 155, 165–166
    and Quaker asylum design, 178, 182, 183, 191
    of steel as construction material, 155, 161
    subjectivity of standards in building construction, 154, 161–162
    in tall timber construction, 163–164, 166–167, 170
SAIC-Frederick, 89, 113n13
Saraiva, Tiago, 7, 12, 23–24, 25, 208, 244
scale/granularity, and gold RM, 82
scanning tunneling microscope (STM), 75
Scattergood, Thomas, 184
Schaffer, Simon, 4, 15
scholarship of things, 3–4
science and technology studies (STS), 6–7, 9, 17–18
Science magazine, 136
scientific investigation, 16–17
scientific surveys (of agriculture), 131–134
security. *See* safety
segregation and white supremacy, in citrus growing communities, 123, 141–142
Senaga, Karen, 12, 21–22
sensory experiences, 20
Shamel, Archibald D., 137–138, 140–141
Shelby, Richard, 64
Sheller, Mimi, 215
Sheu, Jeffrey, 221–222
ship in a bottle metaphor, 109
shutters, on asylum windows, 184–185
skyscrapers. *See* tall timber construction
slow burning mill construction, 153
Smith-Lever Act (1914), 134
soil alkalinity, 48

sports. *See* athletes, performance of; bicycle frames
spreadsheet as bureaucratic document, 95–97
stability, 86–87
standardization, 22–23. *See also* building codes and regulations; bureaucracy, and development of nanosize standard
   and citrus industry, 143–144
   of fish flavor evaluation, 53–54, 57–64
   and materiality, 80–81
   of nanoparticle size measurements, 84–88, 90, 102–103
   of orange orchards, 136–137, 138–140
   and social domination, 143–144
   as social institution, 79–81
standardized scientific objects, 74–75
Star, Susan Leigh, 81–82
Stark, William, 183, 192
steel
   in bicycle frames, 212
   as construction material, 152, 155, 157, 160–161, 165
Steel, Eric, 85
Stephens, Chester "Check," 39, 40
Stora Enso Oyj, 155
STRAL (catfish processing company), 39–40
Subcommittee on the Admission of Light and Air (SALA) (of Building Committee of Friends' Asylum), 26, 177, 187–188, 190–191
Subic, Aleksander J., 206
subject/object distinctions, 17–18, 246
supply and demand, as problematic categories, 8, 10

sustainability in construction, 152, 155, 157, 170, 171

T3 building, Minneapolis, USA, 160
tactile experience
   of Grenfell Cloth, 237
   and window design, 190–191
Taiwan, manufacturing in
   of CFRP, 217–220
   "Made in Taiwan" products, 27, 218, 220–222, 223–226, 229
tallest mass timber buildings, 160, 162–163
tall timber construction, 24–25, 151–171
   fire safety demonstration tests, 166–167
   government support for, 158, 161
   historical significance of, 160
   materials, fabrication of, 155–157, 165
   public perception of, 153, 155, 163, 166
   regulations permitting, 154, 159, 162, 163
   relationship to other material safety standards, 155
   vs. steel and concrete, 155–158, 160–161, 163, 165–166
   types of, 164
tall timber construction, advantages of
   aesthetic aspects, 151–152, 153, 160
   carbon sequestering, 160
   ease and speed of construction, 156, 160, 165
   fire resistance, 153, 158
   sustainability, 152, 160
   tolerances, tightness of, 165–166

tall timber construction,
  disadvantages of
  challenges in success of, contextual, 153–155, 159, 166, 169, 170
  construction sites and fire risk, 168–169
  flammability, 152–153, 161, 162–163
  perception as preindustrial material, 152, 166, 170
  safety, unknowns regarding, 163–164, 170
"tall timber industrial complex," 154
Taniguchi, Norio, 75, 111n2
taste testers, 53–56, 59–60, 62–63
techno-bureaucratic objects, 81–83, 97–107
  and bureaucratic indifference, 97–100
  and bureaucratic mundaneness, 103–107
  and bureaucratic rigidity, 100–103
  and standard-making, 109–110
technological innovation
  cast as sources of betterment/virtue, 13–14
  and national identity, 207–209, 210–211, 227
TED Talks, 151–152
textile production, 1, 252–254. *See also* Grenfell Cloth
Thaysen, A. C., 49
Thornton, William, 193
tilapia, 52
timber industry, government incentives for, 158
Timber Innovation Act (2017), 161
timber plantations, 158
time, organization of in historical narratives, 30, 276

Timmermans, Stefan, 80
titanium, 215
Torero, Jose, 12, 24–25
Torrans, Les, 52
Tour de France, 210–211, 215, 217
tourism, and bicycle racing, 210, 211
traditionalism, 271–272
transmedia stories, 28–29, 245–246, 259–264, 265–266
Treet building, Bergen, Norway, 160
Treves, Frederick, 247
True, Richard, 39
Tuke, Samuel, 181, 182–183, 184, 185–187, 192
Tuke, William, 181
Twitchell, James B., 214
Type X gypsum board, 163

Underwriters Laboratories, 153
Union Cycliste Internationale–International Cycling Union (UCI), 222
United Kingdom
  building codes in, 159
  Grenfell in, 246, 247, 251–252
  York Retreat, 181–184, 189
United States
  building codes in, 159–162
  research funding and incentives for tall timber construction, 158, 161
United States Department of Agriculture (USDA), 46, 59, 122–123
United States Green Building Council, 155
United States Standards Strategy, 91
University of British Columbia, 162
University of California at Riverside, 134

University of Chicago Laboratory
    School, 121, 132
University of Nottingham, 168–169
uplift (by wind loads), 164, 168
utensils used for cooking and eating,
    272–274

values. *See also* aesthetics; country-of-
    origin stereotyping
  construction materials embodying,
    24–25, 167, 170–171
  of materials, 8–14
  and moral treatment of the insane,
    178–179, 180–181, 195–196
Vancouver, BC, Canada, 162–163
van der Ploeg, Martine, 60, 61
Vaux, Roberts, 193
Vinsel, Lee, 10
voluntary consensus standards, 91,
    113–114n29

Wasiak, Patryk, 12, 27, 33n18
Watkins, Henry, 254–255
Watters, Bob, 101
weather. *See also* Grenfell Cloth
  and clothing, 29, 243, 248, 254
  impact on flavors of farm-raised
    catfish, 47, 51
Webber, Herbert John, 134, 135–136
websites
  of cycling communities, 220,
    225–227
  of Grenfell England company,
    262–263
  Haythornthwaite family site,
    259–261
  and transmedia storytelling,
    259–264
Wilier (Italian bicycle manufacturer),
    211

Wilson, David, 212
wind, and clothing, 29, 237, 248
wind loads (in construction), 164,
    168
window glass, 185, 186, 189
window sashes, 178, 184–191
  design of, 184–187, 190
  fabrication of, 188–189, 190, 191
  at York Retreat, 183
wire rope, as construction material,
    154
Wise, M. Norton, 91
Wistar, Thomas, 177, 188, 193
wood as construction material, 24–
    25, 151–171. *See also* tall timber
    construction
  aesthetics of, 151–152, 153, 156, 160
  engineered wood products, 156–157,
    167–168
  and fires in high-rises, 12, 15, 25
  flammability, methods of deterring,
    158
  heavy timber construction, 164
  lightweight timber construction,
    156, 158, 164
  perceived as premodern
    technology, 153, 154, 170
  perception of safety for high-rises,
    155, 165–166
Woolgar, Steve, 82
World Trade Center, 155
World War I, 247–248
wrought iron, structural, 154

Yaneva, Albena, 7, 197n5
Yarnall, Ellis, 193, 194
York Retreat, 181–184, 189

Zeller, Thomas, 208
Zimba, Paul, 63